Spiritual Insights from the New Science is a guide to the deep spiritual wisdom drawn from one of the newest areas of science — the study of complex systems. The author, a former research scientist with over three decades of experience in the field of complexity science, tells her story of being attracted, as a young student, to the study of self-organizing systems where she encountered the strange and beautiful topics of chaos, fractals and other concepts that comprise complexity science. Using the events of her life, she describes lessons drawn from this science that provide insights into not only her own life, but all our lives. These insights show us how to weather the often disruptive events we all experience when growing and changing.

The book goes on to explore, through the unfolding story of the author's life as a practicing scientist, other key concepts from the science of complex systems: cycles and rhythms, attractors and bifurcations, chaos, fractals, self-organization, and emergence. Examples drawn from religious rituals, dance, philosophical teachings, mysticism, native American spirituality, and other sources are used to illustrate how these scientific insights apply to all aspects of life, especially the spiritual. *Spiritual Insights from the New Science* shows the links between this new science and our human spirituality and presents, in engaging, accessible language, the argument that the study of nature can lead to a better understanding of the deepest meaning of our lives.

Spiritual Insights
from the New Science
Complex Systems and Life

Other Titles by the Author

The Gate of Heaven and Other Story Worlds
Belle O' the Waters
Fearless

Spiritual Insights from the New Science

Complex Systems and Life

Raima Larter
Former Professor of Chemistry, School of Science,
Indiana University-Purdue University at Indianapolis, USA

World Scientific

NEW JERSEY • LONDON • SINGAPORE • BEIJING • SHANGHAI • HONG KONG • TAIPEI • CHENNAI • TOKYO

Published by

World Scientific Publishing Co. Pte. Ltd.

5 Toh Tuck Link, Singapore 596224

USA office: 27 Warren Street, Suite 401-402, Hackensack, NJ 07601

UK office: 57 Shelton Street, Covent Garden, London WC2H 9HE

Library of Congress Cataloging-in-Publication Data

Names: Larter, Raima, 1955– author.

Title: Spiritual insights from the new science : complex systems and life / Raima Larter.

Description: Hackensack, New Jersey : World Scientific, 2021. |
    Includes bibliographical references and index.

Identifiers: LCCN 2021003541 | ISBN 9789811232244 (hardcover) |
    ISBN 9789811233821 (paperback) | ISBN 9789811232251 (ebook for institutions) |
    ISBN 9789811232268 (ebook for individuals)

Subjects: LCSH: System theory. | Computational complexity. | Religion and science.

Classification: LCC Q295 .L358 2021 | DDC 201/.65--dc23

LC record available at https://lccn.loc.gov/2021003541

British Library Cataloguing-in-Publication Data

A catalogue record for this book is available from the British Library.

For any available supplementary material, please visit
https://www.worldscientific.com/worldscibooks/10.1142/12158#t=suppl

Desk Editor: Shaun Tan Yi Jie

Typeset by Stallion Press
Email: enquiries@stallionpress.com

Printed in Singapore

For all my teachers

Contents

Introduction

I was a research scientist and professor, immersed in the investigation of complex systems science, when I realized that this science held profound insights for my life. And not just my own life, but everybody's.

This was an area of study that didn't even exist when I entered graduate school in the mid-1970s, but by the time I'd become a faculty member and started to establish my own research program, this "new science", as James Gleick dubbed it in his book *Chaos: The Making of a New Science*, had started to show promise. We soon found it to be a source of tools and analytical methods that could reveal the secret workings of things as tiny as a single cell in our body, or as large and complex as society, a range of study that few areas of science can cover.

Nowadays, this new science is most commonly referred to as complex systems science or the science of complexity. Some of the earlier names for it include chaos theory, bifurcation theory, nonlinear dynamics and the science of emergence.

What I didn't expect when I'd embarked on my chosen field of study several years before, was that it would provide the means to understand what was happening to me on a beautiful September day when anybody else would have said I was suffering a midlife crisis.

It felt like a crisis, for sure, but what I was actually experiencing was growth.

As I considered my own situation, I realized that the science of complex systems can provide key insights into those moments of change and transformation that happen to all of us — and not just to individuals, but to organizations, communities and groups of all kinds, including society as a whole.

In comparing my own experiences to those of others, I quickly realized it wasn't just *my life* that had considerable overlaps with these scientific concepts, it was everybody's life. The personal events that led me to that turning point was just one among many types of events that could be understood through the lens of this new science. Getting tenure, learning a painful truth, receiving a cancer diagnosis, becoming a parent — all these, and more, can lead to the type of growth I was experiencing.

I began to see commonalities between a whole range of transformational events and key concepts from the science of complex systems. In addition to developing insights about the *attractor* and its *bifurcations*, I also began to see how *self-organization*, *cycles*, *chaos*, *fractals* and *emergence* — all concepts from the science of complex systems — could be understood at a deeper level, revealing insights about our spiritual life.

The seven sections of this book consider each of these insights in turn: first from the point of view of how they help us understand, in a scientific sense, complex systems; and, second, by considering their deeper meaning. I will also discuss how mystical and religious experiences sometimes reveal these same truths and how these insights from science are reflected in other types of human activity: dance, ritual, religious tradition, and other ways of knowing.

I've used my own life to illustrate how I came to understand these ideas better, not because I think what happened to me was in any way special or unique, but because my life is the one I know best. I hope describing my experiences in this context might help others see how these same concepts apply to their lives.

It is also my hope that this knowledge, drawn from both a scientific study of nature and other ways of knowing, will help my readers understand and cope with growth and transformational change as they really occur — and to embrace the process of being alive.

Insight 1   The Attractor

I live my life in widening circles
that reach out across the world.
I may not complete this last one
but I give myself to it.

I circle around God, around the primordial tower.
I've been circling for thousands of years
and I still don't know: am I a falcon,
a storm, or a great song?

Rainer Maria Rilke[1]

An attractor is like a valley in a mountainous landscape. We can envision a complex system as a ball moving over this terrain. The ball will eventually roll into the valley, the attractor, and stay there. To fully understand this metaphor, we need to know more about the ball: the complex system.

Complex Systems — A Definition

Complex systems are "smart". This means that they *adapt* when conditions change, rather than merely react to those changes. The attractor can help us understand how a complex system is able to be so adaptable.

A thermometer is not a complex system, since an increase in temperature merely causes liquid to rise up the thermometer's glass tube a certain distance in proportion to that temperature change. The thermometer simply *reacts* to the change. It is not smart enough to adapt to the changing temperature.

Our bodies, of course, *are* smart enough to adapt — within limits. Other complex systems can do this as well. Consider, for example, an ecological system. This collection of plants, animals, microbes, bodies of water, wetlands, deserts, and so on is a system in which the parts are connected and interdependent. An ecological system *can* adapt to temperature changes. It is smarter than a thermometer.

Every time the season changes, as temperatures plummet at the approach of winter or soar when summer arrives, an ecosystem adapts and survives. Even if those changes become long-term or even permanent, an ecological system has ways to adapt. Ecosystems are smart because they can adjust to a change in climate by, for example, shifting the mix of species. In this way, the system as a whole survives. The end result of this adaptation could be a very different ecosystem from the original, though. Individual animals or plants, or even entire species within that system may not make it, but the system as a whole will likely survive. It has adapted.

This is not to say that climate change is "okay". In order for an ecosystem to adjust to something as severe as the warming of the entire planet, it is highly likely that some species will go extinct. One of those species could be our own.

Our immune system is another example of a smart, complex system. It will adapt by ramping up to an activated state when a virus or bacterium enters our body. All systems in the body are adaptable and, thus, complex. Social systems are also considered to be complex systems, since they have the ability to adapt to changing conditions. In fact, social systems, in many ways, act more like a living organism than like a machine. All of these complex systems have one characteristic in common: they are capable of adapting to change, sometimes in creative and unexpected ways.

Nonlinear Dynamics

Nearly the entire history of physical science has been devoted to studying what might be called the *static structure* of matter. Atomic theory, concepts about the nature of molecules and subatomic particles, even the description of the components of life — tissue, cell, cytoplasm — are all ways in which matter has been described as if it is a constant, unchanging thing.

But matter is not static. It is constantly in motion and changes over time. To understand the dynamic nature of matter, the ways in which the parts that make up objects and lifeforms move, interact, and evolve over time, we need non-static theories. Thermodynamics was one early theory developed to describe a certain dynamic aspect of the universe, in this case the flow of energy and heat from one object or region to another. Classical mechanics is another such theory, and quantum mechanics a third.

None of these approaches, though, is adequate for understanding or explaining the behavior of complex systems. This is because complex systems do not obey linear laws and all three of the approaches listed above are restricted to linear equations. When graphed, linear equations yield a straight line. Nonlinear ones, however, are curved. Nonlinearity in the laws that govern complex systems makes them more difficult, but not impossible, to study.

In the early days of science, solving a nonlinear equation took a lot of time and effort, since all calculations were done with pencil and paper. It is only since the advent of the modern computer that we have been able to efficiently study the dynamics of systems that obey nonlinear equations. Complex systems science came on the scene when there were tools available that were up to the task, in the early 1960s when computers became widely available. This is, in fact, the main reason the study of complex systems is referred to as a *new science* — it burst on the scene nearly a half-century after quantum theory, which was itself once referred to as the "new" science. Both of these were considered new sciences for their time because of the way they revolutionized methods used to investigate the natural world.

The Universality of Complex Phenomena

Prior to 1963, the year Edward Lorenz published his seminal paper[2] on chaos in a weather model, virtually no scientific publications or conferences existed on the topic of nonlinear, complex systems. Since that time, the field has exploded, and one of its main features is the large amount of cross-disciplinary collaboration. Biologists share insights with chemists, physicists, mathematicians and medical doctors at these conferences. Social scientists and computer scientists find common ground at the joint meetings.

The reason these cross-disciplinary exchanges are possible is that there is a universality to the phenomena displayed by complex systems. Universality provides a common language that has allowed for fruitful collaborations across disciplines.

One type of universality can be seen in the types of dynamic phenomena exhibited by complex systems. Examples include oscillations, both periodic and chaotic, propagating fronts, spatial patterns and waves — a plethora of different behaviors exhibited by systems as diverse as a collection of inert molecules, interacting populations of predator and prey, aggregating amoeba, flames, fluids — even a dripping faucet.

The universality of complex behavior is the key to our ability to understand so many different types of nonlinear phenomena. Once the complex system is reduced to a set of nonlinear dynamic equations, it doesn't matter if we are studying a chemical reaction in a Petri dish or an isolated island where a population of wolves and moose interact as predator and prey. Math is what unites these two examples.

In my own work, for example, I have been able to apply insights from my lab's study of chemical reactions to processes that occur in the human brain. It turns out that the chemical oscillations we studied in our lab have a great deal of similarity, in a complex systems sense, to the oscillations observed in a human patient in the throes of an epileptic seizure.[3] We discovered this by looking at the dynamic processes occurring in cells known as neurons, important components of brain tissue, and other cells known as glia. Neurons, surrounded by a matrix of glia, are connected in networks, transmitting information from one cell to the next through electrical signals. These information-transfer processes can be described in ways similar to those used to describe chemical systems. This is but one example of the way that universality has helped us to greatly expand our understanding of complex systems.

Another type of universality comes about when transitions known as bifurcations occur. For example, the sequence of events leading up to a transition to chaos is often the same from one system to the next. A very common sequence involves what is known as a period-doubling bifurcation cascade. A bifurcation is basically a fork in the road. We will explore bifurcations in the next chapter and will return to consider this and other aspects of the universality of chaos in Chapter 5.

Another example of universality that we will delve into more deeply in a later section relates to a quantity called the fractal dimension.[4] It is remarkable just how many phenomena in nature can be described using fractal geometry, and even more remarkable that the same fractal dimension is often observed in disparate situations. This particular universality reveals information about the way the fractal object has developed or grown. More on this phenomenon in Chapter 6.

Complicated is not Necessarily Complex

Ordinary systems, while not complex, are not necessarily simple, either. A jet plane, for example, is a very complicated machine, but it is not complex in the way our bodies or an ecological system are. Complex systems seem more "alive" to us than ordinary systems, no matter how complicated those ordinary systems are. Complex systems seem this way because they adjust to change by strengthening or weakening the interactions between different parts of themselves, or even by changing out the parts of which they're made — just like living things do. It is worth noting that many modern systems, including jet planes, now incorporate smart technology, allowing some level of adaptation to changing conditions.

Living things adapt to changes in their environment by adjusting their internal state. Every living organism is, then, a complex system in its own right. Organisms have an internal "force field" that governs the whole range of complex interactions that allow the organism to adapt so as to stay alive and remain healthy.

While your own body has this same type of inner force field — a complex, interconnected set of processes involving proteins, enzymes and other biomolecules — you don't, of course, have conscious control over any of it. However, your body is "smart" enough and can adapt to change as needed, without you being consciously aware of every shift in molecular action or biological function.

The Attractor: Historical Record and Forecasting Tool

Some simpler complex systems, such as chemical reactions or fluids, can be studied in a laboratory and the internal force fields that attract those systems towards a certain class of behaviors can, in fact, be measured. The attractor is a reflection of these internal forces. It is simply a plot of the system's trajectory through state space.

This definition has a lot of probably unfamiliar terms in it (*plot, trajectory, state space*) so we will take the definition apart and consider each term in detail.

The attractor can be visualized by simply graphing, or *plotting*, the system's behavior over time. An example is shown in Figure 1. These graphs help us understand both why the complex system exhibits certain types of behavior and also how the system is able to adapt to changing conditions.

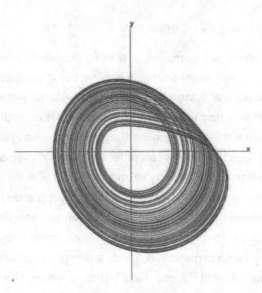

Figure 1: The Rössler Attractor. This diagram shows trajectories for the two variables x and y calculated from the Rössler system as it exhibits chaotic behavior.

The attractor is a visual representation of the way the system's internal force field acts on that system, in much the same way that the earth's orbit is a visual representation of the results of gravity causing our planet to orbit the sun. The orbit is the earth's *trajectory* through space. The elliptical path that the earth follows when it makes its annual trip around the sun is, of course, not a visible path; there is no ellipse or circle in space that some traveling spacecraft might be able to photograph. The orbit is simply a drawing of the trajectory, created by *plotting* on a piece of paper (or in a computer) the places where the earth has been in the 12 preceding months as it traveled around the sun.

The attractor diagram for a complex system is constructed in the same way as a planetary orbit, by plotting the system's behavior over a certain period of time. The plot is not drawn in physical space the way a planet's orbit is, but in a "space" where each point corresponds to precise values of the system's measureable properties.

These measureable properties might be chemical concentrations, populations of animal or plant species, or fluid velocities. Many types of variables, chosen for different types of systems, can be used to visualize the attractor. The axes in this type of space are the values of the variables that define the state of the system, so I will refer to the space as *state space*. Another term used in the complex systems literature for this type of space is *phase space*. This terminology is borrowed from classical mechanics, where a graph of position versus momentum, used to describe trajectories such as planetary orbits, is referred to as a plot in phase space.

Each point in a state space corresponds to a set of properties associated with one state of the system. The attractor is quite literally a picture of the "orbit" of the complex system in state space. It is a visual representation of the history of the complex system, just as the earth's orbit shows its history — where it has traveled around the sun.

While the orbit is a record of the earth's past travels, its most powerful use is as a forecasting tool. The orbit shows the trajectory our planet is expected to follow in the future as it continues to be acted on by the force that governs its motion — in this case, the force of gravity. This forecasting ability is, of course, why space flight is possible. Scientists and engineers can precisely calculate where all the planets and moons in the solar system will be at any time, once their orbits are all known.

The attractor is, similarly, a powerful forecasting device. By observing how the complex system has behaved in the past, we can predict, to some extent, how it will evolve and develop in the future. We can think of it this way: imagine that the complex system has an inner landscape that governs its motion through state space. There may be hills and valleys, ridges and channels in this imaginary landscape. All of these features will determine the pathway that an imaginary river will follow as it flows over the landscape.

The attractor is, then, a map that shows the river's course over this landscape. It shows where water in the river has flowed in the past and can be used to predict where it will go in the future — as long as the attractor retains its shape and the underlying landscape stays the same.

When the attractor bifurcates, though, the landscape abruptly changes and our prediction abilities become much more difficult or even impossible. We will consider the bifurcation phenomenon in more detail in the next chapter.

The calculations used to forecast future behavior of a complex system are not carried out in the same way as they are for planetary motions. This is, again, because complex systems obey nonlinear equations of motion. Complex systems behave in ways that planets do not. Sometimes they circle in orbits through state space that look almost like planetary motion, but at other times the motion can be more elaborate — loops inside of loops or even more complex forms. Occasionally, a complex system exhibits a type of behavior called *chaos*. It is possible to predict some things about a system that is behaving chaotically, but not its precise state at any given moment in time. We will delve more deeply into chaotic behavior in a later chapter.

A Spiritual Attractor?

When I first began to consider the possibility that an attractor was at work organizing my psychological, emotional and spiritual life, I thought about how I could ever know for certain if that were so. I was a scientist and I longed for some hard data to test my idea. I wished there were some kind of spiritual or emotional thermometer I could deploy to measure my internal state. Then, I would plot the data the way I was doing in my lab for our chemical systems.

There were explicit mathematical tests I could apply to see if an attractor really was at work in my life — if only I were able to get a good measurement of my internal state. These tests would answer the question that was starting to take hold: Could it be that an attractor organizes my inner life as it does with the chemical systems that make up my biological body? And, if there is such a thing as a spiritual attractor, does it also arise from the chemical forces powering my body? Or does it originate somewhere else altogether?

Alas, there was no such spiritual thermometer available to test my idea, but I have come to believe, although there is certainly no scientific proof of this, that there is, indeed, an attractor that organizes the complexities of our spiritual lives and gives them form and structure. I found evidence for this idea in surprising places: dance, ritual, the insights of poets and philosophers, and accounts of those who have reported mystical experiences down through the ages.

I knew from my research that an attractor is not imposed on the system from any external source. It's created by the internal workings of the system itself. I also knew that if a system has an attractor, it is likely to be a smart system. It is also likely to be complex, not simple like a thermometer. Anything with interconnected inner working parts — a cell, an organism, a family unit, even a nation — will suffice. Furthermore, just like attractors we can study in the lab, an attractor for any of these systems will be created by the system whose dynamics it governs: by the system itself. Since the complex system I was thinking about was *me*, an individual human being, it must be my *self*, I reasoned, that creates my own attractor.

But I knew that this was only half the truth. If my life created its own attractor, that attractor would, in turn, govern the types of behaviors that I could experience in the course of that life. The attractor would pull me toward certain activities and behaviors, while repelling me from other ones. These influences on my behavior might feel like restrictions. Certain things could even seem unavailable to me.

All this seemed to be very much in line with the way I'd experienced my own life, so I dove deeper into the idea. If the attractor allowed or disallowed certain activities, this would be a reflection of the way my internal river could flow through the various channels available to it. If the inner landscape changed, I might find that things I used to be attracted to would now repel me. It might be that my internal attractor had once allowed certain things, but now it disallowed them — and the allowed or disallowed nature of these activities was not due to a set of rules that other people or groups might decide or vote on. These limits on my life choices were a reflection of some major upheaval in the underlying landscape that had once shaped my life's trajectory. If I tried to force my internal river to flow uphill, it just wouldn't happen.

When I was 33 years old, I received tenure at about the same time my youngest child entered first grade. All of this should have made me happy, but instead I felt as if everything I had worked for was not what I wanted. I saw my life as governed by an inner attractor, and the attractor's name was Science. And it seemed that my attractor had just bifurcated — but what was it urging me toward now?

I had been passionate about my work for decades, drawn to it first as a student. Looking back, I can see evidence even then of the attractor at work on my life choices. One piece of evidence was this: I could get lost in my studies, taking no notice of the passing of time. I remember once, jerking my head up to look around at the darkening room in which I'd been studying. I'd sat down after lunch to work on Chemistry homework, and the clock showed that it was now well past 6:00 pm. More than five hours had elapsed without me being aware of the passage of time.

Mihaly Csikszentmihalyi is widely credited with introducing the idea of a sense of timelessness as proof of positive mental engagement with the task at hand. He gave[5] the process the name "Flow", which seems quite perfect for describing the alignment of one's own personal energy with the flow of the life process around the attractor. I believe that what Csikszentmihalyi means by "Flow" is the sense of being aligned with one's attractor, allowing its tugs, influences and urgings to begin to shape the course of one's life. It's a good feeling, and when you are lucky enough to experience it, pay very close attention, for it is one of the most informative encounters with the attractor you can have.

Science AND Religion — Not Science vs. Religion

When I hit an apparent bifurcation at age 33, I found myself asking, "Is this all there is to my life?" I knew I now wanted something other than what science could provide — but what was it? As I began to consider the possibility that my life, indeed everybody's life, is governed by an attractor and that mine had just bifurcated, I knew I had come upon an insight that would take me far beyond the bounds of science and into the realms of spirituality and religion.

Although the media may portray Science and scientists as anti-God, and the views that make it into the public sphere often make it seem as if belief cannot be reconciled with a commitment to the methods and findings of science, this is simply not true. Recent surveys[6] have shown that the majority — nearly 70% — of scientists are either religious or spiritual, and of the remaining 30% only a small portion are true atheists, the rest being agnostic or simply uninterested in religion, as I was for many years.

However, this same study also showed that most scientists who have a religious or spiritual life are reluctant to speak openly about their beliefs, feeling that to do so would raise questions about the rigor of their work or their commitment to the scientific method. This is understandable given the pseudo-scientific approach of proponents of certain ideas like creationism or intelligent design — ideas that pretend to be science when they're not. The media landscape is dominated by only a very few, loud voices on both sides of the argument, and neither are representative of the thoughts and beliefs of most of us in science. This is unfortunate, not only because it's inaccurate, but because science and religious traditions have much to teach each other.

Science can never prove or disprove the existence of God, and those who claim their data shows that there is no God are stretching science beyond its intellectual limits. This is also true of those who claim the opposite — that they have, somehow, proved the existence of God. Neither of these claims of "proof" is remotely possible.

However, despite the inability of the methods of science to rule on the existence of God or even the relative truth of one set of beliefs or another, I do think we can learn a bit about the nature of the divine world, especially our own spirituality, by deep investigation of the natural world and scientific theories about it. It is in this spirit that I offer these insights, in the hope that they might be of help to all.

Rituals to Evoke the Attractor

Religious traditions of all types are firmly rooted in ritual, and many examples seem to be attempts to evoke a personal experience of the attractor. Joseph Campbell has explained that rituals are the enactment of myth and, therefore, allow us to have a physical experience of the story that informs our religious tradition. As Campbell put it, "Myths are the mental supports of rites; rites the physical enactments of myths."[7] Those who participate in ritual believe that they are physically enacting an underlying spiritual truth, so the ritual is a means by which a believer can more deeply and intimately access that truth.

Rituals that lead to a physical awareness of the attractor are, then, a way to externally enact the internal experience of it. Rituals of circumambulation, i.e., walking around a center point or object endowed with divine significance, seem especially evocative of the attractor, and are widespread across cultures and times, and common in multiple religious traditions.

Islam, for example, has the Hajj, a ritual pilgrimage to the holy city of Mecca. All faithful Muslims hope to make such a pilgrimage one day, and the ritual involves many steps, including one component in which masses of pilgrims walk together in procession around the gold-domed mosque, the Kabah, or symbolic center of the holy city, as shown in Figure 2. This circumambulation of the Kabah is a physical enactment of a spiritual event the pilgrims believe is occurring inside each one of them.

Karen Armstrong, in her book *A History of God*, quotes the late Iranian philosopher Ali Shariati, as he describes the experience of participating in the circumambulation portion of the Hajj: "As you circumambulate and move closer to the Kabah, you feel like a small stream merging with a big river. Carried by a wave you lose touch with the ground. Suddenly, you are floating, carried on by the flood. As you approach the center, the pressure of the crowd squeezes

Figure 2:   Muslim pilgrims circumambulate the Kabah during the Hajj.

you so hard that you are given a new life. You are now part of the People; you are now a Man, alive and eternal."[8]

Shariati, Armstrong says, goes on to explain that the Kabah is like the world's sun whose face attracts each pilgrim into its orbit. If Shariati had known about the attractor, perhaps he would also have described this part of the Hajj as a way for the believer to experience the flow of the trajectories around their own internal attractor.

This circumambulation ritual is a powerful reminder to each pilgrim that he or she is part of a universal system, a small particle in a great river of humanity, flowing around the central organizing point, which is, to the believer, Allah. Far from making each person feel small and insignificant, though, the process provides the believer with a sense of dissolving into the mass of faithful pilgrims as they all dissolve collectively into Allah, becoming one with their God. "This is," according to Shariati, "absolute love at its peak."[8]

Circumambulation rituals abound in most religions, but are especially prominent in Hinduism and Buddhism.[9] In the Hindu traditions, circumambulation often occurs as part of temple puja, or worship, and the object that is circumambulated is either the temple itself or an image of the divine, such as a statue of the god to whom the temple is devoted or a general symbolic image of divinity, such as a linga.

In Buddhism, circumambulation around a stupa is common. The stupa, a sacred container, may have within it a relic of the Buddha or one of his disciples but, more importantly, it is symbolic of that which the Buddhist seeks: the source of enlightenment. As in the Hajj, circumambulation around a stupa is generally a component of a pilgrimage to the holy site where the stupa has been placed. Tsultrim Allione explains it in her book *Women of Wisdom* this way: "By circumambulating or practicing near a great stupa one's mind is tremendously benefited and the merit of the practice is multiplied a thousandfold."[10]

Other rituals which evoke the attractor place the center around which circulation occurs *within us*. Circle or spiral dances in the religious practices of many indigenous people come immediately to mind. Another striking example is the whirling dervish dances characteristic of Islamic mysticism. Created by the Sufi poet Rumi, these dances evoke for the individual dancer an experience of a deep, inner cyclic flow and, simultaneously, allow the dancer to participate in a ritual that he believes is necessary for the fundamental cyclic process of the universe to continue.

The Sufi believe they must whirl in their dance always and for all time, in order for the fundamental spin that is at the core of all life and the world to continue. Some have speculated that this underlying cyclic process the Sufi sense might be spinning atoms or electrons, but I wonder if, perhaps, the dancers are whirling in synchrony with their inner attractor.

Experiencing Your Own Attractor

It is possible to sense at least a little of what the Sufi dancers do by creating a personal spinning ritual. I was taught how to do this by my friend Annie Carpenter, a modern dancer who worked for years with the Martha Graham Company before moving to Indiana where she taught dance and movement meditation. At the time I met her, Annie, now a well-known yoga teacher,[11] was developing mindful movement techniques to help her students achieve greater self-awareness.

She invited me to join her and several others in a Monday morning movement class to teach us how to "spin", a controlled motion she described as analogous to, but much slower than, a figure skater's spin. When done in just the right way, Annie said, spinning induces a meditative state.

We began by standing quietly in our bare feet on the wooden floor of Annie's studio. She instructed us to place our feet parallel to one another, planted on the floor about shoulder width apart, with our hands cupped loosely together and held gently in front of our throats.

Annie recommended cupping the hands lightly enough that a small slice of light was visible between the thumbs. If you bring your gaze to this small spot of light, she said, and look steadily at this focal point, you are less likely to feel dizzy when beginning the spin.

After a few moments of standing quietly, we began slowly rotating the right foot outward by sliding it on the floor one quarter turn, then immediately bringing the left foot around to meet it, returning both feet to their original parallel positions, our bodies now rotated by 90 degrees. This sequence was repeated again and again: slide right foot, then left, slowly turning to the right.

I had to be reminded more than a few times to keep my gaze fixed on the sliver of light through my cupped hands. I also had to be reminded to keep both feet on the floor, to slide them along the surface rather than lifting them with each turn. This would help me feel more stable, Annie said, as she encouraged us to turn, turn, turn, at an exceedingly slow pace.

After a while I started to get the hang of it. I began to trust that I wouldn't fall. I found I could increase my speed a little. Following her instructions, I let my hands float outward to my sides. I could still visualize, in my mind, the sliver of light that had been there when my hands were cupped.

I started to spin faster, and after some time something remarkable happened.

That spot of light, real or remembered, began to direct me toward a central spot deep within my psyche or soul. As my physical body spun, I realized it was not just my body, but also my inner self that was rotating around a deep central point.

With this very simple method, one I could do anywhere with a bit of space, I had located the center of my own attractor.

Stopping the spin must be done carefully, Annie cautioned. It is important to slow down *gradually* to retain one's balance. With practice and repetition, she said, you might even be able to carry the experience of feeling completely centered around with you for many hours after you stop the spinning meditation.

Annie taught this technique as a simple, personal way of finding and reconnecting with the center we all have within us, but easily lose track of when we get caught up in our hectic lives. She describes her own experience of finding the center through spinning this way: "When I first begin the spin, I am aware that *I* am turning and the *room* is stationary. But, gradually, I begin to sense that the *room* is turning and *I* am stationary. When this happens, the turning motion begins to become effortless and I feel no dizziness or out-of-control feeling. All the concerns of the day, the chaotic and unpredictable nature of life, melts away as I find the center of the turning motion. I visualize the center as an axle passing through my body, connecting the earth below my feet to the sky directly above my head. For me, spinning is all about finding order in the chaos."[12]

Annie has discovered a modern way to bring some ancient wisdom into our lives. Twyla Nitsch, a 20th-century Native American teacher from the Seneca nation, Clan of the Wolf, teaches that everyone has a "beginning place" and "a circle within the mind", a phrase that is highly evocative of the attractor. Nitsch's advice for dealing with what I have described as a bifurcational time is quite similar to Annie's advice and involves staying focused on the center from which the attractor originally springs: "Through life it's important that we all make our circle within the mind, and that we stay within that circle, because that's our sacred space."[13]

Rituals can, then, help us connect with our own personal attractor, but sometimes an awareness of this attractor simply "breaks through", without anything in particular being done to induce that awareness. Kathleen Raine, a self-described "nature poet" who says "...nature poetry is not what we write about nature but what nature tells to us," describes an experience she once had, induced by simply looking at a flower, that is striking in its evocation of the inner attractor: "I could apprehend it as a simple essence of formal structure and dynamic process. This dynamic form was, as it seemed, of a spiritual not a material order; of a finer matter, or of matter itself perceived as spirit. There was nothing emotional about this experience which was, on the contrary, an almost mathematical apprehension of a complex and organized whole."[14]

When I first read this poetic description of a slow circulation of vital current, apprehended almost mathematically, I was struck by what seemed to be a vision of an attractor in a spiritual state space. Raine seems to have been in touch with this attractor for a long time and to have found her way to its center point at a very early age. She writes of her experience as a child, sitting on a small grass-covered ledge that was her "secret shrine": "This soft seat of fine grass, and rock rising above it, covered with such abundance of fern as I had never seen, was, for me, that focus and hub of the world that human beings are always looking for. …But here I had it, and sat like a bird on her nest, secure, unseen, part of the distance, with the world, day and night, wind and light, revolving round me in the sky."[14]

If we are between stages in our life and our attractor is bifurcating, rituals such as the spinning one recommended by Annie Carpenter, and other techniques I will describe later, can be especially centering. If we happen to be in a chaotic stage (and, I suspect, this is likely to be the normal stage of a mature life), rituals that help connect us with the stability of the underlying attractor can be especially important. I will have much more to say about the true nature of chaos in a later chapter, especially the amazing things that happen to our attractor when we enter the chaotic realm. Trying to *stop* the chaos will not work; rather, we must learn to enjoy the ride and trust that the attractor, which continues to exist and work its influence on our lives even when it bifurcates, will continue to stabilize our existence.

Endnotes

1. "Ich lebe mein Leben…/I live my life in widening…" by Rainer Maria Rilke, copyright © 1966 by Anita Barrows & Joanna Macy; from RILKE'S BOOK OF HOURS: LOVE POEMS TO GOD by Rainer Maria Rilke, translated by Anita Barrows and Joanna Macy. Used by permission of Riverhead, an imprint of Penguin Publishing Group, a division of Penguin Random House LLC. All rights reserved.

2. Edward N. Lorenz, "Deterministic Nonperiodic Flow," *Journal of Atmospheric Science*, Vol. 20, p. 130 (1963).

3. (a) Brent Speelman, Raima Larter and Robert M. Worth, "A Coupled ODE Lattice Model for the Simulation of Epileptic Seizures," *Chaos*, Vol. 9, pp. 795–804 (1999).

    (b) "Just Slow Down: There's Trouble in Store when Neurons get Too Speedy," *New Scientist*, Vol. 163, No. 2204, p. 16 (1999).

    (c) "Better Living through Chaos," *The Economist*, Vol. 352, No. 8137, pp. 89–90 (1999).

4. (a) Michael Barnsley, *Fractals Everywhere*, Academic Press, New York (1988).

    (b) Benoit B. Mandelbrot, *The Fractal Geometry of Nature*, W.H. Freeman and Co., San Francisco (1983).

5. Mihaly Csikszentmihalyi, *Flow: The Psychology of Optimal Experience*, 1st Edition, Harper Perennial Modern Classics, New York (2008).
6. Elaine Howard Ecklund, *Science vs. Religion: What Scientists Really Think*, Oxford University Press, New York (2010).
7. Joseph Campbell, *Myths to Live By*, Penguin Books, New York (1972), p. 45.
8. Karen Armstrong, *A History of God*, Knopf, New York (1993), p. 156.
9. Willard G. Oxtoby and Roy C. Amore, *World Religions: Eastern Traditions*, 3rd Edition, Oxford University Press, New York (2010).
10. Tsultrim Allione, *Women of Wisdom*, Snow Lion, Revised & Enlarged Edition, New York (2000).
11. Learn more about Annie's work at her website, anniecarpenter.com.
12. Personal interview with Annie Carpenter.
13. Anne Bancroft, *Weavers of Wisdom: Women Mystics of the Twentieth Century*, Penguin Books, London (1989), p. 40.
14. Kathleen Raine, *Farewell Happy Fields*, Hamish Hamilton, London (1974).

Insight
2

Bifurcation

Two roads diverged in a wood, and I —
I took the one less traveled by,
And that has made all the difference.

Robert Frost[1]

A bifurcation is essentially a fork in the road. Imagine traveling along in complete darkness. If you couldn't see that you'd arrived at a branch point, how would you react? Perhaps you wouldn't react at all if you had no idea you needed to make a decision. Not seeing that the new path differs from the one you had been traveling on will have consequences. We respond to the fork in the road differently if we know it's there, so learning about attractors and bifurcations can help us navigate what might, otherwise, be a tumultuous transition.

Bifurcation of an attractor can be both painful and exhilarating and it can happen to organizations and individuals alike. The characteristics are the same whether it is an individual or a family, church, business — even a nation — that is undergoing it. Sometimes precipitated by an unexpected or even traumatic event, such as a cancer diagnosis, but equally as often the result of natural growth, the individual or group undergoing this type of transformation suddenly know that the life they once led, individually or collectively, is now over. The river that once carried that life along so smoothly has now encountered a barrier that wasn't there before and a new path must be chosen or the change acknowledged.

The whole world has changed and the things the individual or group members say to themselves, and to others, reflect this certain knowledge: "Nothing will ever, ever be the same," some say. Others express the sense that "we live in a whole new world now," or "there's no going back." All they know for certain is that something fundamental about their former life has undergone a complete transformation.

The World will Never be the Same

I didn't fully appreciate the enormity of the change that had occurred as a result of my personal bifurcation until I saw it happen[2] again to all of us in the days immediately following September 11, 2001. "The world has changed," people said. "Nothing is the way we thought it was," and "we will never see things the same way from now on." The exact nature of that change has been repeatedly discussed and is still, decades later, the subject of debate, but there is unanimous agreement that the world was dramatically changed by that event. You can hear it in our language when the "pre-911 world" is contrasted with the "post-911 world", and we all understand exactly what that means. Although just one country, the US, was attacked, human society as a whole encountered a bifurcation point that day. The underlying landscape of humanity's collective attractor shifted, and we have struggled ever since to adjust to the new reality.

It is not just disturbing events such as terrorist attacks that can lead to bifurcations of our collective social attractor. Bifurcations also happen when technological breakthroughs occur. The invention of the printing press, the first radio transmission, the first human flight, the launching of the Internet — all these events completely transformed the landscape of our society,[3] changing the evolution of culture in ways that could not have been foreseen before each breakthrough.

Even though we might label a terrorist attack as a "bad" bifurcation and the appearance of new technology as "good", the changes that occur as a result of either are equally traumatic. The invention of the printing press, for example, led to the relatively unrestricted circulation of ideas — something that had happened very slowly before, if at all. This altered the structure of society in deep ways and it was not an easy transition for the people of the time.

Knowledge shifted from the hands of political and religious authorities to the common person, but this was resisted by those in power.

We have had over five hundred years to adjust to the idea of mass communication through books and newspapers and we now deem the effects of the printing press to be "positive", but we need only to look at the turmoil surrounding the free flow of information through the Internet to see the stresses and strains on society that can result from even those events we might deem to be "good" bifurcations.

Growth by bifurcation in individuals sometimes occurs in response to a crisis or an unforeseen shattering event such as a serious illness or the death of a close friend or family member, but "good" causes are also possible — the birth of a child, for instance, is clearly a bifurcation in the new parent's life, as any mother or father can attest. More often, though, I suspect bifurcations occur just because we reach a certain age: 13 or 18 — or even 33.

In my own case, completing my 33rd year on this earth might have been the sole reason for the bifurcation that occurred that beautiful September afternoon — or it might have been the fact that I had just received tenure. Or, it might have been that my youngest child had just entered first grade, changing my own sense of identity in a way I could not fully appreciate until many years later when he left home for college, launching me into yet another bifurcation.

Or it might have been all of these.

For a long time, I attributed the break to the sudden insight I'd had from taking a short self-guided quiz in a book about childhood sexual abuse.[4] The quiz was meant to diagnose lingering effects of the abuse and I had answered "yes" to every question. One in particular, though, had hit me hard: *Do you use work or achievements to compensate for inadequate feelings in other parts of your life?*

I'd been sitting on the front porch as I took that quiz and when I reached this question, I could no longer sit still and ran to the mailbox at the end of the driveway — as if the emptiness welling up inside could be filled by a piece of mail. It couldn't, of course, and I felt my life crack open as everything I'd worked for was called into question.

In hindsight, I believe that quiz merely put a match to the dry tinder that had piled up when all those other changes in my life began accumulating — tenure, an emptying nest, and reaching my mid-30s. All of these, together, triggered the bifurcation, but the true cause lay within me.

The existence of multiple causes and triggers can even be considered a piece of evidence that the change we are dealing with is the far-reaching transformational event called bifurcation. It also seems to me that this may be one of the reasons for the arguments about what actually "caused" 911, and whether it should have been foreseen or predicted. Bifurcations are often over-determined, meaning there are multiple causes, so it may be the same with the September 11th attacks — there were multiple causes. We have become used to thinking that society works in mechanical, linear terms — one cause leads to one effect — but this is not the way living things grow and change. And human society resembles a living organism more than it does a machine.

Bifurcation of one's life attractor is no small thing. It is wrenching and far-reaching, and not at all pleasant while it is going on. It's a process akin to metamorphosis, and we can hardly imagine how that particular life transition must feel to the poor caterpillar whose body suddenly begins to break down to its very molecules. However, the end result of the caterpillar's disorienting growth, a process that must feel more like destruction than growth, is quite literally a new life. And when we successfully navigate the bifurcation of our own life's attractor, we, too, will find ourselves living an entirely new life.

The necessity to make decisions when we encounter one of these bifurcation points is often resisted. More often than not, I suspect, we don't even realize a decision is needed. We are so caught up in the events of the transition that, as we approach the fork in the road, we close our eyes to the fact that the path we are traveling on is ending and new paths are opening in front of us. We ought to notice this, ought to pay attention. We need to make a choice about which new path to take, after all — instead, we march right on through, out into the tall grass and weeds or straight up that steep hill, often oblivious to the fact that our life terrain has shifted and we must now find and follow a new course.

Before we know it, our insistence on keeping things the way they were, sticking to the original path — the one that no longer exists — has left us far from any path at all. We may be unable to find our way back to either the old life, the original path, or to locate our new life — one of the forks in the road we can barely see. In fact, if we ever *were* able to find our way back to where the original path was, we would find that the old path is probably no longer such an attractive option or may not even exist any longer.

This can be very confusing, especially if something as unsettling and disturbing as an unwanted bit of information about our past, a medical diagnosis we don't want, or the death of someone close has occurred. Now what do we do? What path do we follow? We can sit at the bifurcation point, not knowing how to move, as long as we refuse to see that the landscape has shifted, our attractor has bifurcated and we are now called to choose one of the new paths that had not existed before our life terrain changed so dramatically.

As discussed in Chapter 1, the ability to make forecasts about future behavior is one of the more powerful aspects of the attractor. When a bifurcation occurs, though, the very shape of the attractor changes, along with the underlying landscape. Imagine how all our forecasting ability about the earth's future position in its orbit around the sun would be thrown into disarray if the orbit suddenly changed from an ellipse into a knot or another complicated shape.

This, in fact, is exactly what happens to an attractor when the very dramatic internal change known as a bifurcation occurs. The "orbit" is no longer what it was. When a planet is pulled into an orbit by a gravitational field, the planet moves toward its proper position and velocity. When an attractor acts on a complex system, though, it pulls it toward certain *behaviors* — not just toward certain *places*, which is what the gravitational force of the sun does to an orbiting planet. When an attractor bifurcates, reflecting an abrupt change in the underlying force field and landscape, the *behaviors* that attract our system can shift dramatically — and this can be very confusing, if we are the complex system in question.

I had been following a path when this occurred to me and it felt as if the ground beneath my feet buckled and reformed itself. The topography was inverted. Valleys became hills, mountains turned into ravines — and the internal river that had carried me over this inner topography could no longer find the channel through which it had once flowed.

At one point, it seemed as if this internal river ran headlong into a high, wide cliff, a barrier that had not been there before. I fought against it for a while, refusing to admit that the landscape had changed, but the river continued to flow, threatening to crush me against that immovable wall. What I didn't understand at the time was that if I stopped struggling, the water would rise and lift me over that wall. Once I allowed myself to be carried over this barrier, I was deposited, with little effort, onto fresh soil — and a new life.

Why is it Called Bifurcation?

If you travel through areas of the United States covered with pine forests, you will undoubtedly see swaths of brown trees that have been killed by a caterpillar known as the spruce budworm. There are both eastern and western varieties; the western species goes by the name *Choristoneura freemani* and the eastern one is *Choristoneura fumiferna*. Each of these is a moth when in adult form, but it is the larval form, a small caterpillar called a budworm, that is particularly destructive.[5] While it is true that each budworm undergoes a bifurcation in its transition from caterpillar to moth, there is another type of bifurcation that occurs for the full population of budworms.

The life cycle of the budworm/moth is completed in one year. Wildlife biologists have worked hard to understand what causes the larval budworm population to sometimes swell to great numbers. The budworms eat fresh pine needles on conifer trees, so there is a lot of food available in a forest. It may seem like they could eat and grow indefinitely, but a forest does have a finite carrying capacity — a maximum amount of food available to the budworms. In addition, the caterpillars are preyed upon by birds, so the budworm's numbers are limited by both predation and the finite food supply.[6]

When the budworms reach the adult moth stage, they reproduce by laying many eggs, each of which will become a new caterpillar. This increases the population, of course, while the other two forces — limited food supply and predators — decrease it. Both the predator population and the finite carrying capacity of the food supply serve as checks on unfettered population growth — but they also lead to a bifurcation in the type of attractor wildlife biologists have found for this species.

For low values of the carrying capacity — call it K — the population of spruce budworms remains at a single low value. As the carrying capacity increases through a certain critical value — call it K_{crit} — the complex system describing the budworm/pine forest/bird ecosystem goes through a bifurcation. At this point, three possible populations of the spruce budworm become available, and one of the new ones is much larger than the original population. When K becomes greater than K_{crit}, the population of budworms essentially explodes, leading to massive deforestation.

This ecosystem shows us where the word "bifurcation" comes from. A graph of the spruce budworm population as a function of the carrying capacity looks like a pitchfork with one line to the left of K_{crit} and three prongs branching out of it to the right (see Figure 3). The single line to the left of the critical point and the three prongs to the right are the available spruce budworm populations. The special value of the carrying capacity, K_{crit}, is a bifurcation point for this ecosystem since the graph forks, or bifurcates, at this critical point.

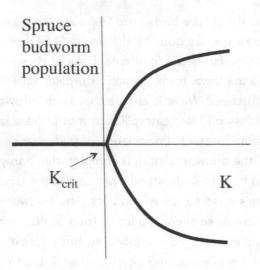

Spruce
budworm
population

K_{crit}　　　K

Figure 3: Pitchfork bifurcation in the spruce budworm model.

The language carries over to the underlying attractor. The pitchfork-shaped diagram is a reflection of a bifurcation in the attractor itself. When the carrying capacity reaches and subsequently exceeds the critical value, the attractor for the ecosystem is said to bifurcate. At this critical value, a single budworm population becomes three new possible populations — one lower than the original, one the same as the original, and one much larger than the original. The middle one is, as it turns out, not stable, so it appears as if the single population branches, or bifurcates, into two new ones. This type of situation is referred to as multi-stability,[7] since there are multiple stable states available to this ecosystem for high values of the carrying capacity.

It may seem as if the value of the carrying capacity K is the sole cause of a bifurcation, but the critical point K_{crit} depends on the other parameters in the system — the amount of predators around and the birth rate of budworms — so, as usual, the bifurcation is over-determined. There are multiple reasons why this ecosystem can undergo a bifurcation, a statement that is true for all bifurcations of complex systems.

The Power of Metaphor

After the day in my 33rd year when I encountered an apparent bifurcation in my own personal attractor, I struggled mightily to figure out what it meant. Did it mean I should no longer work as a scientist? I had loved doing science, but now I began to see that I loved doing other things as well. One of these other things I loved was writing.

The simplest interpretation, one I considered frequently, was that I had just discovered that my choice of profession had been wrong. Let me be clear: I no longer think that choosing science as a career was wrong, but these black and white ways of thinking are typical of a time of bifurcation.

My thoughts kept returning to one increasingly obsessive idea: maybe I should quit my job and become a writer. It made sense in light of the ideas about attractors I was beginning to develop. What had attracted me before, now repelled. The repetitive sameness of my days, weeks and semesters that had once provided comfort, now seemed almost like a spiritual death. I was attracted to new and inexplicable things — like writing poems, which I didn't know how to do. Whenever I would allow my thoughts to get too close to the possibility that the organizing force of my life had changed, though, a great upswell of protest would hit me from within. There were so many reasons to reject this idea, not the least of which was the fact that to follow these crazy impulses seemed to mean both financial and career suicide.

Besides the impractical aspects of this sudden desire to change my life in a way that made no sense, I was plagued by doubts about whether I was misinterpreting. What about the sense of the unseen force I'd felt all these years, that "attractor" that pulled me deeper and deeper into the study of Science? It had seemed so real. Why couldn't I feel it anymore? How could I, or anyone, know if the desires we feel are signs of an attractor at work in our lives, or if we are just being selfish?

When I tried to apply the insights from the science I had been studying my entire career to my current situation, the abstract nature of my emotional and spiritual life stymied me. A literal approach didn't seem to work, since how would I measure the spiritual or emotional "variables" that were relevant, how would I make plots, look for bifurcations, and so forth? Even though there was no way to know if an attractor literally existed for these more ephemeral parts of my life, I finally relaxed into the idea of applying the concepts from my science in a metaphorical way to my inner life.

The metaphor of an inner landscape, going through upheavals and causing my attractor to bifurcate, was helpful to me because it provided a visual aid, a picture, (or, better yet, a video) of my personal experience of spiritual growth. A metaphor helped to highlight those things that were most central to my experience, and I began to see that what was most important to me was also of universal importance. This was a very new mode of thought for me, since it involved none of my scientific, analytical ways of thinking. What I was beginning to discover was how poetry or myth can get at the essential truth better than a mere re-telling of a historical event.

As Karen Armstrong notes in *A History of God*, "The one [history] describes what has happened, the other [poetry and myth] what might. Hence poetry is somewhat more philosophic and serious than history, for poetry speaks of what is universal, history of what is particular."[8] Similarly, a literal interpretation of the attractor concept applied to one's own life would graph an individual life's trajectory, recording what has happened — but a metaphorical interpretation would tell what might happen, to *anybody*, which is much more useful. A metaphor would illuminate the possibilities, not just be a mere recounting of events already lived.

Metaphors are much more powerful than we often realize. Gregory and Mary Catherine Bateson point out that metaphor is not just "pretty poetry... but is in fact the logic upon which the biological world has been built, the main characteristic and organizing glue of the world of mental process."[9] As George Lakoff and Mark Johnson explain in their book *Metaphors We Live By*, "The essence of metaphor is understanding and experiencing one kind of thing in terms of another."[10] We use metaphor all the time in ordinary speech when discussing emotional concepts that cannot be described literally. As Lakoff and Johnson explain, concepts such as emotions cannot be clearly delineated in a *direct* way, and so must be comprehended *indirectly*, i.e., by metaphor. And what is true for our emotional life is even more true for our spiritual life.

To illustrate the power of metaphor, Lakoff and Johnson explain that love has been variously described as a physical force ("They gravitated toward each other" or "I could feel the electricity between us"), as madness ("I'm crazy for him") or even as war ("He won her hand" or "She relentlessly pursued him") because we cannot say directly, that is, literally, what love *is*. Likewise, I could not say directly what a spiritual attractor literally is, and I found a metaphorical description to be much more useful.

A metaphorical description or understanding is not "less than" a literal one. It can, and often does, lead to a deeper and more helpful understanding of the concept than a mere literal description would allow. I was especially persuaded by a story,[11] recounted by Lakoff and Johnson, of a student whose native language was not English and who had come to understand the phrase "the solution of my problems" with a chemical metaphor in mind. The student understood this familiar English phrase to mean a boiling, smoking liquid containing all his problems. Some of those problems were dissolved in the liquid, some suspended as solid precipitates in a solution containing, perhaps, catalysts to speed the dissolution of some problems or cause the precipitation of others.

The student's metaphorical understanding provides a *very* different, and powerful, view of "solving our problems" than the one we might arrive at using a different metaphorical understanding of finding the solution to a problem, such as one drawn from solving puzzles or equations.

The student's metaphor implies that the problems don't disappear when they are solved or, as the student envisioned it, *dissolved*. The problems are still there, just not in their original solid form and, furthermore, there is always the possibility of a problem reappearing when you are in the process of solving, or dissolving, another problem. In many ways, the student's metaphorical understanding is much closer to the truth of what "solving" problems actually means, since we never *really* solve our problems — we simply transform them or, if we are lucky or insightful, actually convert those problems into opportunities.

The choice of a proper metaphor can give us deep insight into complex issues, such as those that confront us when our life has suddenly changed dramatically. Keeping in mind the metaphorical image of a landscape undergoing occasional terrain-changing earthquakes was of tremendous help to me when the organizing force of my own life began to create dramatic changes in my interior life. This mental image, of a spiritual landscape going through upheavals that changed the entire character of the terrain around me, forcing my life path into new trajectories, was one I returned to again and again.

Even though I had begun to see the wisdom of the attractor metaphor for my own life, I could not relax. I knew that it is not usually the case that only one bifurcation happens. Instead, a sequence of bifurcation events, rather than a single one, typically takes most attractors through a series of transformational changes — so what was going to happen next? I knew that many systems undergo the same, or a very similar, sequence of bifurcations as they evolve, so I wondered if that would be the case for my life as well.

I figured I was in some kind of cyclic, or oscillatory, state since the repetitive cycles that I was sensing with greater degrees of discomfort certainly seemed to fit this interpretation. I knew, though, that when a cyclic attractor bifurcates, one of three things usually happens: a more complex but still cyclic and repetitive attractor could replace the first. Another possibility is that equilibrium, in other words death, is reached. This didn't seem to be the case, but the third possibility (and this was the part that scared me) is that chaos ensues. Several more years were to go by before I would learn, though, that the chaotic stage of our attractor holds some of the most profound of all spiritual lessons, and I would find, with great surprise, that I would eventually welcome the appearance of chaos in my life since it inevitably brings with it deep learning and growth that are unachievable any other way.

The Wisdom of the Mystics

Whether you subscribe to a particular religious tradition or to none, whether you are filled with doubt or with certainty about the existence of a divine power, you and I are alike this way: we will all be faced with myriad decisions in our lives, turning points at which we seem to be called to move in a different direction from the path we have been following during the previous months and years. Whether we believe that someone — or something — is "watching over us" or whether we believe that nothing in the universe beyond our friends and family is particularly interested in the details of our personal life, all of us will, at times, experience the overwhelming power of personal growth. At these times we sense an overpowering or irresistible force propelling us forward and compelling us to make some changes. This can be both exhilarating and frightening.

The metaphor of a governing attractor may provide comfort at such times of profound growth. We might want to think of this metaphor as a valuable lesson provided solely by the most modern new science, but this is simply not true. Spiritual thinkers down through the ages have understood the significance of what I call the attractor, although they didn't call it that, of course. But they also didn't need science to intuit its truth. Consider, for example, Simone Weil's evocative description of the inner spiritual landscape: "He whose soul remains forever turned in the direction of God...finds himself nailed to the very center of the universe. It is the true center, it is not in the middle, it is beyond space and time, it is God. In a dimension which does not belong to space, which is not time, which is indeed quite a different dimension, this nail has pierced a hole through all creation, through the thickness of the screen which separates the soul from God."[12]

It is hard for me to read this description without thinking of it in terms of an attractor. Simone Weil was a contemporary writer (1909–1943), but the ideas she wrote about have been around for a very long time. The Greek philosopher Plotinus (~240 CE) liked to compare "the One to the point at the center of a circle, which contained the possibility of all future circles that could derive from it."[13]

Bruno Borchert also wrote about Plotinus's evocative concept of the center. Borchet explains that the midpoint was, for Plotinus, in each thing the same creative Spirit: "Everything flows from this Spirit and everything has a tendency to become one with the One again. This tendency is a cosmic stream, an impulse, a cosmic eros, which Plotinus calls insight."[14]

These writers, as well as those who preserve and protect rituals and other sacred traditions that seem to draw on an intuitive knowledge of the attractor, have been trying to guide us toward this truth for a long time. Many of these attempts to put into words the profound experience of a spiritual attractor can be quite poetic. One of my favorites is Nicholas of Cusa (1401–1464) who wrote, centuries ago: "Every spirit finds it sweet to gravitate continually to the very center of life. For a persistent and continual gravitation toward life is the elemental element of ever-growing happiness. What do the living seek but life? The existent but existence? What does love seek but to be loved?"[15]

Nicholas apparently had, at some point, a profound mystical[16] experience that changed the whole course of his life and inspired all his later work. In this experience he came to understand that we all have within us "the very center of life", which he identified as God. Nicholas was an accomplished scientist, mathematician and philosopher. He contributed to the development of geometry, calculus, and physics. He was also a Canon in the Roman Catholic Church and a contemplative monk who wrote about his insights achieved through what we might today call meditation. His mystically achieved teachings were influential in the origin and study of ideas about infinity, infinitesimals, and the like — ideas that ultimately contributed to the development of calculus.

Borchert argues that it was Nicholas, not Copernicus, who lived a century later, who first postulated that the earth orbited around the sun. As we have all been taught, this hypothesis was later embraced not only by Copernicus but also Galileo, who became the first and most famous scientific martyr for defending this scientific fact to the Church, which didn't want it to be true.

Nicholas's insight about a heliocentric earth-sun system was apparently due to an experience that led him to an understanding of the human spirit *not* as the center of the universe, but as orbiting around another center. "We find the middle of the world not on earth, but in God,"[17] according to Nicholas. It seems supremely ironic that Galileo would be arrested by and excommunicated from the Church for championing an idea that seems to have originated in an event that Nicholas, a church official, considered to be a peak religious experience.

To Nicholas, the lessons of this experience for our personal life and relationship with God took precedence over any astronomical insight it might provide about the solar system in which we live. Bruno Borchert's explanation of Nicholas's philosophy is especially evocative of the attractor image: "The human being is the midpoint of creation insofar as he or she is a microcosm. According to Nicholas of Cusa, this does not imply that our dwelling place, Earth, is central, too. Earth orbits round another focal point, the Sun, and our solar system is one of many such systems in an unending universe. Men and women are centers only to the extent that they are specially rooted in the center of all: 'We find the middle of the world not on earth, but in God.' And God is central in each part of the universe, down to the tiniest atom."[17]

Although Nicholas was a high-ranking priest in the Church he *did* suffer for espousing these ideas and was called to Rome numerous times to explain himself. It is likely that the idea of "God in the center of us" was as alarming to the church fathers in Nicholas's day as it was in Galileo's — and, I dare say, as it still is today. After all, if God is *within* us, who needs a priest or pope to mediate?

The forces that nudge us toward spiritual growth often seem profoundly subtle and terribly unclear; we are not always sure what we are being called to do or to change at those times we are feeling the influence of the attractor. Sometimes it seems that we must fight these forces in order to move forward; at other times it seems we must give in to their influence and allow ourselves to be carried forward with no clear idea of where we're being taken.

Some have suggested it is fear that causes us to struggle against the forces propelling us to change our life. But what is it we are afraid of? Stephen Levine proposes[18] that it is a sudden knowledge of our own immensity that causes fear. Levine says that at times of grief, sorrow, illness, death and despair we are brought into contact with the greater reality of which we are a part, and this awareness can be very frightening to us. I agree: at times of bifurcation, it becomes much easier to sense the vast spiritual landscape that has been there all along.

Finding the proper metaphor can create a new reality for us as powerful as the one experienced by the student who found a way to visualize problem solving as something besides solving puzzles. A metaphorical understanding of life under the influence of an attractor which bifurcates, an attractor that dies and is reborn in new form, and does this over and over and over again, can similarly cast a whole new light of understanding on the true nature of growth and change. It is possible to accept the notion of an attractor as a bit of wisdom revealed solely by the scientific study of the physical world, without attributing any divine significance whatsoever to it. It is possible, of course, but I wonder how much of the motivation for this skepticism is mostly about the need to control the uncontrollable.

Endnotes

1. Robert Frost, "The Road Not Taken," from the book THE POETRY OF ROBERT FROST edited by Edward Conner Lathem. Copyright © 1969 by Henry Holt and Company. Copyright © 1936 by Robert Frost. Copyright © 1964 by Lesley Frost Ballantine. Reprinted by permission of Henry Holt and Company. All Rights Reserved.

2. Some of the material in this section appeared earlier, in slightly different form, in this article: Raima Larter, "Life Lessons from the Newest Science: Bifurcation," *Noetic Sciences Review*, No. 59, pp. 22–27 (2002).

3. Phyllis Tickle, *The Great Emergence: How Christianity is Changing and Why*, Baker Books, Grand Rapids, Michigan (2012); This book looks at a several-hundred year sweep of history to explore how one religion changed when technological breakthroughs occurred.

4. Ellen Bass and Laura Davis, *The Courage to Heal: A Guide for Women Survivors of Child Sexual Abuse*, Harper & Row, New York (1988).

5. This website contains a lot of information about the spruce budworm and its devastating effects on western coniferous forests: https://csfs.colostate.edu/forest-management/common-forest-insects-diseases/western-spruce-budworm/.

6. James D. Murray, *Mathematical Biology*, Springer-Verlag, New York (1990).

7. I have listed here a few general textbooks that explain the phenomenon of multi-stability as well as other key concepts from nonlinear science; all are aimed at an audience of undergraduate science students or beginning·graduate students.

   (a) Gregoire Nicolis, *Introduction to Nonlinear Science*, Cambridge University Press, Cambridge (1995).

   (b) Steven Strogatz, *Nonlinear Dynamics and Chaos*, Addison-Wesley, Massachusetts (1994).

   (c) Arthur Winfree, *The Geometry of Biological Time*, Springer, New York (1980).

8. Karen Armstrong, *A History of God*, Knopf, New York (1993), p. 37.

9. Gregory Bateson and Mary Catherine Bateson, *Angels Fear*, MacMillan, New York (1987).

10. George Lakoff and Mark Johnson, *Metaphors We Live By*, University of Chicago Press, Chicago (1980).

11. Lakoff and Johnson, p. 143.

12. Simone Weil, *Waiting on God*, Routledge Revivals, Taylor & Francis, London © 2018; reproduced with permission of the Licensor through PLSclear.

13. Karen Armstrong, *A History of God*, Knopf, New York, pp. 102–103 (1993).

14. Material excerpted from *Mysticism: Its History and Challenge* by Bruno Borchert, English translation © 1994 Samuel Weiser, Inc., p. 160, used with permission from Red Wheel/Weiser, LLC Newburyport, Massachusetts, www.redwheelweiser.com.

15. James Francis Yockey, *Meditations with Nicholas of Cusa*, Bear and Company, New Mexico (1987); permission from Inner Traditions/Bear and Company to use this and other excerpts is gratefully acknowledged.

16. The word "mysticism" may seem to be synonymous with magic or the occult, but it is a well-known and thoroughly studied type of religious experience. For an excellent overview, see: Evelyn Underhill, *Mysticism: The Nature and Development of Spiritual Consciousness*, 12th Edition, Oneworld Publications, London (1993).

17. Material excerpted from *Mysticism: Its History and Challenge* by Bruno Borchert, English translation © 1994 Samuel Weiser, Inc., p. 258, used with permission from Red Wheel/Weiser, LLC Newburyport, Massachusetts, www.redwheelweiser.com.

18. Stephen Levine, *Guided Meditations, Explorations, and Healings*, Doubleday, New York (1991).

Insight

3

Self-Organization

> *A vast similitude interlocks all,*
> *All spheres, grown, ungrown, small, large, suns,*
> *moons, planets...*
>
> Walt Whitman,[1] *"On the Beach Alone at Night"*

A Chemical Reaction that Organizes Itself

Beyond the ability to adapt to a changing environment, complex systems also exhibit a characteristic behavior known as self-organization. This type of behavior can take the form of spontaneous spatial patterns that appear in an initially uniform medium. Another way self-organization happens is when oscillations or cycles arise in an otherwise quiescent system. Even sometimes apparently disorganized behavior that is actually quite organized can occur. This latter type of self-organization is known as chaos, and will be explored more thoroughly in a later chapter.

Self-organized behavior is just that: it is organized by the system itself, not imposed on it from the outside. Imagine if a slab of cookie dough could spontaneously divide itself into gingerbread man shapes. This cannot, of course, happen, since an external force — the cook — must use a cookie cutter to carve patterns in the dough. A self-organizing system, on the other hand, creates patterns all on its own, without any external help.

The first time I saw an example of self-organization was in college. One day, I attended a lecture and demonstration. A slight, bearded man stood at the front of the room, pouring a clear blue liquid into a large flask. I found an open seat at the back and slid in to watch and listen.

The man, who I later learned was Richard Field, a new Chemistry professor at the nearby University of Montana, switched on a stirring mechanism and the blue liquid began to swirl in the flask. As he began his lecture about things I didn't fully understand, filled with words that fascinated me — inorganic reactions, chemical kinetics, catalysts — I tried to concentrate on what he was saying, but I kept getting distracted by that blue liquid, swirling furiously in the tall, conical flask. I stopped listening altogether when a few streaks of red appeared in the blue. Before the red streaks had traveled even once around its perimeter, the entire contents of the flask blinked to red.

Gasps rippled across the room. Dr. Field, a wiry man with a wry smile, stopped talking. The red color flashed back to blue, eliciting a louder reaction from the crowd. Field smiled broadly, like a magician playing to his audience. As we watched together, amazed and murmuring, the liquid in the flask began an alternating display of color: first red, then blue, then red again — oscillating between one color and the other every few seconds.

"It's called self-organization," Field said. The name, he explained, comes from the fact that the chemical reaction *organizes itself* into orderly patterns, this one a pattern in time, an alternating display of two colors. The periodic flashes of red followed by blue then red again didn't come from anything outside the beaker, he said, but from interactions between the molecules in the flask.

Although I'm certain I didn't understand this at the time, I learned later that one key component of the red/blue oscillating reaction — known formally as the Belousov-Zhabotinsky (BZ) reaction,[2] named after its Russian discoverers — is a catalyst molecule whose color is red in one form and blue in another. The color changes occur as the catalyst cycles between its two states, while facilitating the conversion of reactants in the flask into the final products.

By the end of Field's lecture, the chemical reaction had run its course and the red and blue flashes that had made the flask seem almost alive had slowed and, finally, ceased altogether. As the others filed out of the room, I made my way to the front and bent down to peer into the flask. Now filled with a murky purplish-colored liquid, it no longer looked the least bit alive.

In the days and weeks that followed, I couldn't stop thinking about that lecture. *What made the colors change like that?* Nothing I had ever encountered in my growing-up years — except, perhaps, the bubbling mud pots and geysers in nearby Yellowstone Park — had intrigued me the way Field's demonstration of the BZ reaction had.

Later that semester, I attended another lecture, one of a campus series. This particular talk was given by Stanley Miller of the University of Chicago. He spoke to a packed auditorium about his experiments carried out in the early 1950s with Harold Urey of the University of California–Berkeley. The Miller-Urey experiment,[3] as it is now known, attempted to reproduce conditions present on the early earth at the moment life first appeared. Miller and Urey had mixed sulfuric acid with several other chemicals to simulate the early ocean, then placed the mixture into a large flask pumped full of methane and carbon dioxide. The sulfuric acid mimicked the runoff from volcanoes, while the gases were the likely components of the earth's early atmosphere. Finally, they shot electrical charges through the entire apparatus to simulate the action of lightning storms.

When Miller and Urey and their students inspected the contents of the flask after the discharges were turned off, they found something quite unexpected: amino acids. These molecules, when chained together, form proteins, like those in our bodies. The simulated lightning had apparently caused the atoms in the carbon dioxide, methane and sulfuric acid molecules to rearrange themselves into amino acids: the very building blocks of life. Miller had been very careful when, at the end of his lecture, the inevitable question came from a member of the audience: "Dr. Miller, do you think you actually *created life* in that flask?"

And, of course, he had not, he explained. Those molecules were not *alive*, he said, the way we normally define life, but it was an extraordinary result, nonetheless. The fact that amino acids — the basic components of proteins — were formed, as opposed to, say, the building blocks of plastic, was astounding to Miller and his colleagues. And to me.

As I left the lecture hall that night, swirling in a crowd of students and professors into the cold Montana evening, under a blanket of stars that were always brilliant in that high-mountain town, my brain lit up with a thousand questions: *When does a soup of molecules pass from being just a bunch of chemicals to being alive? Did Miller and Urey's experiment have any relationship to that red- and blue-flashing flask I had seen earlier? Could a solution of amino acids, or any chemicals, keep on changing, self-organizing — somehow — into a life form?*

The molecules in Miller and Urey's flask were not alive, and neither were the ones flashing red and blue in Richard Field's demonstration — but the juxta-position of those two lectures ignited a series of questions that I found I could not turn away from. Dr. Field's explanation of the oscillating color changes as an example of self-organization intrigued me. *If a simple inorganic chemical reaction could organize itself into something as spectacular as a flashing light display,* I asked myself, *what might a soup of amino acids do?* The fact that the mysterious process already had a name — self-organization — suggested to me that these scientists, the ones who presumed they understood this phe-nomenon well enough to *name* it, knew things I wanted and needed to know.

Flocking, Stampeding, and Murmuration

It turned out that I had seen a particularly spectacular demonstration of self-organization before I even knew that's what it's called. I actually heard it before I saw it: a distant rumble, somewhat like thunder, rolling across the foothills toward the pickup truck that my father had just brought to a sudden stop beside the highway. I was about 12 years old and had been sitting in the back of the truck with my sisters and cousins as my father drove us across the wide Lost River Valley in Idaho, near our grandparents' ranch. My dad slammed on the brakes and hopped out of the cab, one gloved hand pointing toward the foothills, the apparent source of the rumbling sound.

There was nothing much to see except the usual: a wide expanse of sage-brush and dry ground. My father often did this sort of thing, bringing the truck to a rapid halt and pointing toward the hills. Generally it meant that he had spotted a deer or elk or even a curly-horned sheep, but this time I couldn't see anything.

The rumble got louder, and then I saw them: a large herd of antelope, crest-ing the hill, stampeding toward us in a cloud of dust. There were hundreds, maybe thousands, of them, all moving as one unit. They swerved, first to the right, hundreds of them, then the left, following a pathway that only they seemed able to sense.

The antelope herd we saw that day may have been running from a predator, since there are coyotes and even grizzly bears in those hills, but no one really knows why herds behave this way. Stampeding is an example of self-organized behavior and is just one of many such examples displayed by different types of species ranging from mammals (like the antelope) to schools of fish, flocks of birds and swarms of insects.

Self-organized behavior like flocking[4] occurs because interactions between the individuals in these systems allow for the emergence of stable group behaviors that are more organized than we might have otherwise predicted, like the beautiful swerving dance of the antelope herd. These emergent phenomena, as they are called, are not something that a single individual in the group is capable of achieving — they are group phenomena, properties of the collective whole.

We have all seen this type of behavior when a flock of birds begins to gather in trees and atop fence posts, lining up in rows on overhead wires. Soon, they take off in a large group, the entire flock swooping this way, then that, always as a unit. If you are very lucky, you might have seen a particularly large group forming elaborate patterns of black speckles in the sky. Videos have appeared online of this phenomenon among starlings; it's called murmuration and is as if the entire flock had simultaneously decided to sketch out a drawing in the air.

And, yet, as far as we know there is no decision. No one bird is in charge, yet the flock, as a collective entity, organizes itself into beautiful patterns. Each bird takes note of its neighbor's position and direction of motion and adjusts accordingly, but nobody is the leader. No head bird tells the entire flock which way to go, yet they move as a unit. The emergence of this group pattern is a characteristic feature of a self-organizing system, and it's not just birds that do it: fish and swarms of insects and even crowds of people do, too.

Notice how your attention shifts to nearby individuals the next time you are in a large crowd of people in a shopping mall or, perhaps, trying to exit a sports stadium or large gathering. It is the movements of your nearest neighbors that largely determine which direction you will go, and also where the crowd will move to next.

One time I experienced this for myself was at the inauguration of the new US President in Washington, DC in January of 2009. This large gathering — more than a million people on the national mall — could have taken a tragic turn since so many people were jostling for position. Scientists who have studied the self-organizing behavior of crowds of people have placed cameras overhead in gathering places and gathered data about their movements. They observed motions strikingly similar to the movements of not only birds and fish, but inanimate systems as well, such as water flowing through tubes.

The data were, in fact, used to devise the crowd control system that was put into place on the mall that day in 2009, using strategically placed barricades and funneling mechanisms.

This same approach has been used[5] in Mecca during the Hajj, another large gathering of people that had previously been marked by dangerous and tragic stampedes until the crowd control system, based on the theory of self-organization, was implemented.

As mentioned earlier, universality is another defining characteristic of complex systems. The identity of the parts that comprise the system is not particularly important — it doesn't matter if it's birds or fish or people that are flocking. What's important is the quality and characteristics of the interactions between the individuals that make up the group. It's these interactions, or relationships, that lead to the emergence of the intricate and, at times, beautiful patterns in a self-organizing system.

The Origin of Life

One of the courses I taught in my years as Chemistry professor was entitled *The World of Chemistry*. Aimed at non-science majors, it focused on the pervasive influence of chemistry in the world around us — food, medicines, materials, etc. — and I enjoyed it immensely.

One of my students, Rosie, came to my office one day to ask some questions, and I ended up telling her about self-organizing chemical reactions. That was when she stumped me with a question: "Do you think that it was self-organization rather than God that made us?"

Even though I was surprised by Rosie's question, it wasn't totally unexpected. Within a few days of our first class meeting, she had let me know that she was "born again", as she called it. Scripture quotations were scattered throughout her conversation wherever she could fit them in and she talked about her "personal Savior", "the Lord", and "our heavenly Father", as if she were discussing a relative or a neighbor.

I have always found it difficult to respond to students like Rosie, some of whom are quite rattled when their classes seem to call into question everything they've been taught. In this case, though, I told Rosie that I actually didn't know the answer to her question.

After all, I had the same question myself. Are we here only because matter is capable of self-organizing into elaborate forms which might even, under certain conditions, take on properties that cross the boundary into the form we would call life? When Rosie first asked that, I saw it as an either-or question. *Either* a god made us *or* we are elaborate examples of self-organized matter that has crossed over that hard-to-define boundary between non-life and life. At the time Rosie asked her question, I couldn't see that *both might be true*: perhaps self-organization is merely the mechanism by which the divine brings forth life *from itself*.

In other words, perhaps God self-organizes and we are the result.

Here I need to interject a statement about definitions. This word, "God", the one that scared me so much the day Rosie sat in my office, is the source of endless arguments and misunderstandings in the world largely because we all mean different things when we say it. Even those who belong to the same religious denomination often mean different things when they use the word. We all have our own personal definition of it, including those who say "there is no God." After all, it is necessary to define what you are denying exists.

Some people take the opposite position, claiming that "there is a God and, furthermore, science proves it." One such claim goes by the name intelligent design.[6] Setting aside the obvious problem here — that science could never prove or disprove something like this — self-organization, which is a scientific fact, shows that arguments like those propounded in intelligent design theory are not based on sound science. The fatal flaw in the "theory" of intelligent design, which is not a theory at all, but simply the latest incarnation of creationism (in other words a religious belief) is its beginning assertion. This assertion claims that life is simply too complex to have arisen spontaneously and an external designer is necessary to explain this complexity.

This statement, always presented as if its validity is self-evident, is not supported by what we now know about self-organizing systems. Much data exist, and solid scientific theories have been developed to explain these data, that show that very complex structures can and do arise spontaneously in self-organizing systems, without benefit of any external designer, or even an external organizing force. While the fact of self-organization doesn't imply (or, conversely, argue against) the existence of a divine designer or creator, it doesn't prove or disprove one, either. But, then, any true scientific theory never would — which is why creationism, in the guise of intelligent design, is not science.

If I were to say "self-organization is the mechanism by which the divine brings forth life from itself," I am making a statement about a religious or spiritual belief, not a statement of scientific fact. The important point is that believing in the sacredness of creation does not require restricting our belief to a divinity who is out there directing the building of the universe like some sort of cosmic construction supervisor.

There are many ways to create things. One way is certainly with your hands or with tools, bringing together starting materials and making something new — a painting, a sculpture or an engine, say. But this is only one of many ways to create.

We can create a dance by moving the body in rhythmic and beautiful ways.

We can create music by opening our mouths and singing.

We can create babies (or some of us can, anyway) with our own bodies.

If we can create in all these ways, why do we assume that the creator of the universe cannot?

This may seem like pagan heresy to some, and I've seen similar theological positions dismissed as pantheism or panentheism,[7] but I'm not making an argument here for one set of religious beliefs over another. If it's pagan or pantheistic to see God this way, so be it.

Religions that require the dismissal of solid scientific evidence are bound to lose adherents since dismissing the insights of science requires us to be less than the incredible species we are: people with brains who have learned how to investigate the most detailed features of the world we live in. I also find it quite hypocritical to dismiss the insights of science regarding geological history or evolution while taking full advantage of the advances that science has made possible, such as the latest medical treatments or new technologies that improve our abilities to communicate with each other or travel around the world. It seems to me that those who dismiss science when it becomes inconvenient or difficult to reconcile with their religious beliefs need to think a little more deeply about where they draw the line in sticking to their convictions.

Similarly, those who dismiss religion as misguided and mock the belief in God as something akin to believing in a flying spaghetti monster seem to have only one rather caricatured view of religion in mind. It is quite possible to both believe in God and fully accept the findings of science. In fact, for some of us, the more we learn about the world around us through scientific study, the more we accumulate evidence that this world we live in is not random and meaningless — far from it.

To me, one of the most inspiring lessons we can learn from the study of self-organization is that matter itself has a propensity for life. It is as if the atoms and molecules of which the world is formed are poised to spring into life and this readiness for life is exhibited in the temporal and spatial patterns we observe in self-organized, but non-living, matter. Whether the self-organization phenomenon occurred in the prebiotic world and carried those proto-biochemicals forward to the point that they actually did spring to life, we will never know for sure. Self-organization *might* be the mechanism, but all we can ever do with hypothesized mechanisms is rule them out. It is impossible to prove that any particular scenario was the one that created the situation we now see millions of years later.

It doesn't seem like life is *currently* arising from inert matter, but we have no way of ruling this out, either. The more we understand the processes of self-organization, the more likely it will be that, someday, a scientist will succeed in coaxing a test tube of self-organizing molecules over the life/non-life boundary. Scientists and bioengineers are intensely involved in trying to do this very thing today. Some have already created complex structures that self-replicate[8] and show certain abilities to adapt to change, just like a living thing. Are these alive? Nobody who has assembled these complexes has claimed they are, but many who are carrying out these experiments are doing so because they want to understand the essential features that separate life from non-life, and they fully believe that the best way to understand something is to make it.

Finally, the fact that self-organization exists doesn't explain *why* it exists. Indeed, we have no idea *why* any of the laws of nature are the way they are. As always, questions of why lead us, quickly, into the realm of questions science cannot answer. After many years of grappling with Rosie's questions, I still find self-organization to be a fascinating phenomenon, but far from a satisfying explanation for something as mysterious as the origin of life itself. Self-organization may have played a role in amplifying the initially small numbers of complex molecules that first formed on the prebiotic earth and it may even have had something to do with how those complex molecules joined together to form living things. I find this a far more satisfying theory of the *mechanism* by which life emerged from inert matter than the non-theory of creationism; belief in a Creator still leaves plenty of room for science to fill in the details about how Creation actually occurred — or, more precisely, is continuing to occur, since clearly the universe we live in is not a static, unchanging entity that, once created, remains the same for eternity.

Many of the most striking examples of continuous creation going on in our universe come from scientific study. The Hubble space telescope has, for example, provided thousands of jaw-dropping photos of stars being born and galaxies forming. Creation doesn't occur only in the heavens, though. Around us, in our own world, new species are arising and old ones are changing and evolving in front of our eyes, an awe-inspiring sight for those who care to look.

The Self-Organizing Slime Mold

I was privileged to witness one truly stupendous example of continuous creation at a scientific conference I attended one year. One of the speakers showed a remarkable video of a strange organism, the slime mold. Known more properly as *Dictyostelium discoideum* (or *Dicty* as it's affectionately known by people who study it) it is, despite its slimy-sounding name, not at all unappealing.

This beautiful bright orange organism that can be found growing on wet logs in forests provides a fascinating illustration of the self-organization phenomenon in biology. The reason the video was shown at our conference was that *Dicty* self-organizes[9] in exactly the same way the BZ reaction, a purely inorganic chemical reaction, does.

The basic feature of a self-organizing system is not whether it's sentient or even alive, but is, rather, its habit of producing patterns — either patterns in time, like the flashing colors in the BZ reaction, or patterns in the spatial realm, like flocking birds. The same system can exhibit both kinds of patterning under different conditions; the BZ reaction that Richard Field used in his demonstration had been set up to produce patterns in time — flashes of blue and red.

Under other conditions, though, this same chemical system can produce elaborate spatial patterns. If the same chemicals that produce red and blue oscillations in the whirling flask are poured into a shallow container such as a Petri dish, and left to sit undisturbed, rotating spiral bands of red and blue will spontaneously appear, as shown in Figure 4. Eventually, those bands will begin to travel across the dish in waves, forming beautiful patterns of counter-rotating spirals or concentric circles.[10]

I was on my way to the lecture hall when I ran into Otto Rössler, a well-known investigator[11] of self-organizing systems, who was arriving late to the meeting. He asked if he had missed anything interesting so I told him about the video we had seen. The film began with what I thought was an overhead shot of a Petri dish filled with BZ reaction mixture: lots of counter-rotating spirals moving slowly across its surface (see Figure 5).

The only difference from the many previous videos we had watched at the conference was that these spirals were black and white, rather than red and

Figure 4: The BZ reaction produces moving waves of alternating red and blue when left undisturbed in a Petri dish. If well-stirred, the color of the whole flask alternates, oscillating between red and blue.

Figure 5: Aggregating slime mold cells during aggregation phase.

blue. I assumed that meant the investigators had not used a color camera, but when the center part of one of the spirals began to clump, then rise up from the surface of the Petri dish, I knew something else was going on.

The speaker then explained that we were not, in fact, watching the BZ reaction but, instead, millions of Dicty cells streaming along spiral pathways toward aggregating centers. Figure 6 shows the process that begins when the amoebae are starved. When the colony of cells is deprived of food and water, a distress signal goes out from these spiraling centers. The video, in fact, showed the beginning of a slime mold evacuation process.

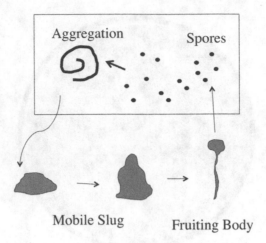

Figure 6: Life cycle of *Dictyostelium discoideum* showing transition from unicellular amoeba form to multicellular organism. The fruiting body form releases spores that give rise to more amoebae.

I explained to Otto how the clump at the center of the spiral kept getting bigger, looking somewhat like a time-lapse movie of a fleshy plant sprouting up from the surface. But, then, it did a very un-plant-like thing: it turned around and *looked at* the camera.

It was both spooky and awe-inspiring. All along, I'd thought I was looking at a video of BZ waves when, in reality, it was a living organism behaving in precisely the same manner. "Ah. I see what you mean," Otto said. "You see the Creator at work, yes?"

And that was exactly it: I had seen the Creator at work.

The Second Law

Even as a beginning student of science I wondered if the laws of thermodynamics,[12] particularly the Second Law, somehow did not apply to living things — but this didn't seem possible. If living things are composed of matter, which they are, they ought to be subject to the same physical principles that all matter is — and the laws of thermodynamics would most certainly apply.

The first of the laws of thermodynamics is simply the law of conservation of energy. It says that the total amount of energy cannot change. This is based on the assumption that there is a fixed amount of energy in the universe. The First Law says that physical processes can only convert this energy from one form to another, not create more.

The Second Law, famous for causing great confusion among science students, can also be stated quite simply: all spontaneous processes result in an increase of disorder. A spontaneous process is one that occurs naturally, without the imposition of an external force or driver. Stated the other way around, the Second Law says that all natural processes decrease the amount of order in the universe.

Thermodynamics was developed in the late 19th century by scientists concerned with the efficiency of engines, the exciting new technology of the time. The reason these scientists, such as J. Willard Gibbs (1839–1903), were so focused on spontaneity, order, disorder, and so on, is that they were trying to understand how to maximize the amount of work that can be extracted from fuel by an engine. The way the Second Law is stated shows this overarching concern. If a fuel is burned in an engine and we wonder how we might get the most out of the energy that the burning fuel releases, the Second Law tells us how to proceed: among all the processes that might occur, the one that maximizes the disorder is the one that will occur spontaneously and without any need for additional injections of energy. Imagine a billowing cloud of ash and waste products swirling away from the burning fuel — this is how disorder increases.

Although the Second Law was developed to understand how engines work and design better ones, Gibbs showed that it applies to all physical processes. Some people have erroneously claimed that living things violate this fundamental law of nature, and since, as a young student, I had considered this possibility myself, I can see why one might think this. After all, if thermodynamics demands that spontaneous processes maximize disorder, then how can an organism, a highly ordered and exquisite construction, appear spontaneously? However, it is simply not true that life falls outside the realm of the laws of thermodynamics — the Second Law *does* apply (as do all the other parts of thermodynamics) and is, in fact, precisely the reason systems self-organize. Because of this fact, it may turn out to be the case that the Second Law is actually *necessary* for the existence of life.

This counter-intuitive insight was made by the late physical chemist Ilya Prigogine and is, in fact, the basis of the Nobel Prize[13] he won for its discovery. The work for which the Nobel citation was given is highly mathematical, so it's not surprising that many people misunderstand it. The upshot of Prigogine's insight is that life is like a whirlpool, or eddy, in a stream. Eddies form when a flowing stream gets caught in an irregular channel. The force of the water in the main channel causes this portion of the stream to whirl, creating a pocket of order that would not exist if the stream stopped flowing.

Prigogine said that living things are like an eddy in a stream of energy flowing from the sun to the earth. Both the water whirlpool and the one envisioned in the stream of light from the sun is a local pocket of order that allows the stream as a whole to maximize what Prigogine called entropy production — the creation of disorder on a grand scale. The maximization of entropy production occurs precisely because the system is governed by the Second Law of Thermodynamics. Prigogine called the whirlpool a dissipative structure, since it is created when the system dissipates excess energy by creating form. The whirlpool exists because the Second Law requires it.

We can think of this excess energy as equivalent to the exhaust that comes out of an engine; indeed, our own bodies would cease to function were it not for our ability to extract material and energy from food, sunlight and air, convert that matter and energy into new tissues and organs, then exhaust the waste products into our environment.

Although Prigogine and his co-workers showed that the Second Law of Thermodynamics explains the self-organizational ability of some systems, we still don't know exactly which systems can exhibit self-organization and which cannot. The designation of some as self-organizing, while applicable to all living things for sure, is also an apt label for many systems we would not normally consider to be alive, such as an oscillating chemical reaction like the BZ.

One of the results of Prigogine's and others' work in this realm is the blurring of the line that separates life from non-life. It is no longer so clear what should be considered as living, since many of the properties formerly associated only with living things have now been observed and studied in inanimate systems such as fluids, chemical reactions and even computer simulations. Perhaps the property we call life does not appear suddenly when a certain level of complexity is achieved but, rather, imbues all of matter at different levels or degrees.

Degrees of Integration

One of the more striking aspects of self-organizing matter is that it displays movement towards greater integration. When atoms bond together to form molecules, such as when two hydrogen atoms and one oxygen atom form water, the resultant properties of the whole are not easily extrapolated from the properties of the parts. The atoms have become integrated into a whole, a molecule, which is more impressive and complex than the parts of which it is made. Similarly, a large collection of molecules has properties that are not exhibited by single ones. Physical chemists have a pithy saying that captures this point well: a single water molecule cannot be wet. Only a collection of water molecules can possess the property we call wetness.

Molecules formed from atoms are actually examples of equilibrium structures. There is a technical difference between equilibrium structures and dissipative structures; equilibrium structures can form in closed systems and are not fed by an external source of energy or matter as is required for dissipative structures.

Small molecules, like amino acids, when connected together become larger molecules, such as proteins, that characterize all life forms on our planet. A great leap in the degree of integration of matter (see Figure 7) is taken when the simplest form of life, a single-celled organism, forms from these larger molecules. Oily molecules called lipids form the boundary, the cell membrane, of this little life, protecting its interior from the harsh environment surrounding it. Within the lipid membrane is protoplasm, consisting of water and electrolytes

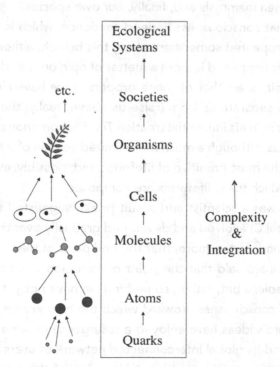

Figure 7: Complexity and integration increase vertically, from subatomic particles such as quarks, to atoms, molecules, and eventually living things.

and laced throughout with complex structures composed of proteins of many different varieties as well as a host of other substances and structures. Even the simplest form of life is a complicated and impressive, highly integrated entity, each part working perfectly in concert with the others, producing an efficient combination which is self-sustaining, self-replicating and enormously stable.

Some forms of single-celled life have persisted for millions of years relatively unchanged. They have survived by taking in matter from the surrounding sea-water, turning it into fresh new structures needed to sustain their small life form, and, finally, undergoing cell division to complete their short life cycle. Other single-celled organisms which arose at the beginning of life on the earth underwent an even more impressive integration by banding together to form colonies which, eventually, evolved into multicellular organisms. This is the point at which the fossil record really begins;[14] very little is known about the evolution of life before the appearance of colonies, but their advent ushered in an explosion in the diversity of lifeforms and the appearance of many branches of the now-familiar evolutionary tree. One such branch led to the development of vertebrates, then mammals and, finally, our own species.

We assume that consciousness (indeed sentience, which is the knowledge that one exists) appeared somewhere along this branch, although the precise point at which this happened is more a matter of opinion than data. Consciousness, like life itself, is another of those properties we have clung to as ours, properties which separate us from (raise us above, really) the lower animals. Teilhard de Chardin in his influential treatise *The Phenomenon of Man*[15] argued that consciousness, although a much more limited version of it than our own, is present in even the most primitive of lifeforms and, possibly, even in the inanimate matter of which these lifeforms are composed.

Teilhard, who was a scientist and Jesuit priest, suggested that Consciousness (with a capital C) evolved and developed gradually over the course of the evolution of life and, furthermore, that the end point of this evolution has not yet occurred. Teilhard said that the point of evolution was not to create the species *Homo sapiens* but, rather, something we have not yet experienced: a global collective consciousness toward which the human race, indeed all life, is moving. Teilhard's ideas have enjoyed a resurgence of fame in recent years as the internet and its global interconnected network of users have raised the possibility in some people's minds that his vision of the future of humanity might not be so far off the mark.

The transition from colonies of single cells to multicellular organisms is a key turning point in the evolution of life and is a stage that is worth taking a second look at. An important distinction between colonies and true multicellular organisms is that the cells in a colony are all identical, whereas those in the multicellular organism have taken on specialized roles, such as blood, skin and brain cells. The step from colonies of identical organisms to multicellular organisms was particularly arduous and occurred over an exceedingly long period of time in comparison to the relative rapidity of subsequent evolutionary steps. Once the jump to multicellularity, i.e., specialization, had been achieved, the dam broke and a flood of diverse life forms, unprecedented and yet to be repeated, spilled out over the oceans of the earth. The differentiation of what were originally all identical cells into an array of specialized cell types is a very obscure process that happened long ago, but it is an event that holds an important lesson for our own time and our own species.

Becoming a Collective

We might ask the question: Does self-organization just occur automatically or is there some role for choice or free will? Maybe self-organizing molecules have no free will, but what about more complex systems — like families, businesses or nations? What about individuals? At what level of organization does free will in a self-organized system come into play?

Perhaps, many millions of years ago, some little single-celled organisms came together, maybe through a mechanism like that which brings slime mold amoebae together, forming colonies that gradually, over many more millions of years, became multicellular organisms like fish. Consider those individuals that, at the end of this process, changed from an independent life form to a mere cell in a body. If those little cells had any idea what was going to happen to them when they first aggregated into colonies, would they have done it?

The slime mold cells seem perfectly comfortable with going back and forth between an individual existence and one as a member of a greater whole, but there is no guarantee when you band together that you will ever be free to break apart again. Would we be so calm if self-organization were happening to us? This is an important question — because, for all we know, self-organization just might be happening to us and our world *right now*.

Just try to imagine the decisions we might be asked to make if we were one of those little cells on the early earth, struggling with the choice of giving up some (or maybe all) of our own individuality for the great unknown of colony membership, whatever that might mean. Accepting a role as a member of this new and improved multicellular form of life would be a very difficult decision to make and I wonder how many of us would go willingly. After all, if a colony of single-celled organisms is tossed upon some sharp rocks at the edge of the sea, the group may be broken apart, the community scattered, but each individual cell would still be able to live on its own as it returned to the simple existence of single-celled life.

But, if a group of individual amoebae had changed into a true multicellular organism in which the individuality of each cell is lost, those individuals will most likely lose the ability to survive on their own as they become skin cells, heart cells and bone cells. Those little cells become *dependent on* the life of the new organism in a way that they weren't before as isolated individuals. If this new life form, a fish or a person, perhaps, were tossed upon these rocks, the entire organism could die, even if only a few cells were actually injured or destroyed in the accident. And if that happens, the individual cell's lifeline would be cut off, and they would die, too.

Joining a colony that might evolve, take on the form of a whole new integrated organism, a new type of life, seems very risky indeed for the individual. We become dependent on the whole in a way we never were before — but, in doing so, we also open ourselves to entire new worlds of experience. Experiences so new that it is impossible for one in their single-celled state to even begin to imagine life as part of a multicellular entity.

Perhaps this is the key lesson of self-organization. When a system self-organizes to a higher degree of complexity, the individual units become so integrated into the more complex new form that they may lose the ability to exist alone. To take such a step (if we even have a choice) requires trust in a future that does not yet exist and whose form is so dramatically different from our current reality nobody would be able to predict even its vague outlines.

If we were one of these early single-celled organisms, how would we feel if some of our friends set off on a journey to the realm of multicellularity? Suppose they became first sponges, then fish, evolving further into mammals, people, civilized human beings. Suppose, further, that some of these pioneers had the ability to return to the world of the single-celled beings, telling of strange and wonderful new experiences awaiting their friends on the other side.

I imagine that they might tell fantastic stories of crawling up out of the water, walking upright over the land, learning to pull together materials and make tools, constructing cities and airplanes and spaceships. They might even tell about the time they had flown that spaceship and walked upon the surface of the moon. If we were still content in our existence as little single-celled organisms, not much bigger than a slime mold amoeba, we wouldn't even know what the moon is. We wouldn't have a clue what our friends were talking about and might even shrug off their fantastic stories as delusional. Why would anyone *want* to self-organize if you lose not only your individuality but, apparently, your sanity as well?

Self-organization can be wonderful and terrible, all at the same time. A unit (or individual) that is part of a self-organizing system will most assuredly feel its current identity threatened as a new, more complex entity, takes shape. Relationships among members of a group or organization that may be undergoing self-organization will change. Chains of command may break or re-form in new and unexpected ways. There is no obvious assurance at such a time that the end result will be better than what existed before — but there is also no reason to believe that any one individual has the power to stop the force which is moving the system as a whole toward a new self-organized state.

Self-Organization in Human Society

If we think of those systems of which each of us, as individuals, is a member, what would we do if we found ourselves in such a situation? How would we react when our family or business — or nation — seems caught up in a self-organization event? And how do we recognize self-organizational change?

First, self-organization is definitely *not* a top-down affair. It is the very essence of grassroots-directed change. Self-organization happens when the parts of a system begin interacting in different ways. The whole system is involved and it is impossible to tell who is in charge, because *nobody* is. In a family or organization or nation, then, one can be assured that it is *not* self-organization if change starts at the top or is under the tight control of a small group.

I've noticed that folks who think about the theory of organizations and management of groups of people have been attracted to the concept of self-organizing systems.[16] The motivation appears to be to devise ways to, somehow, use the insights about self-organization to control or even change the functioning of groups of people. While I certainly share the urge to learn more about how self-organization works, and fully understand the desire to apply insights about this from the non-human world to human systems, I always cringe a bit when I see these attempts. The reason is simple: the person who, as a member of an organization, tries to manage that very organization using insights about self-organizing systems is forgetting that the system functions at a higher level of organization than a single person — even the manager.

Self-organization occurs at the level of the collective, and the forces that govern a collective cannot be controlled by one person any more than the movement of a flock of birds can be changed by one or two lone outliers. Individuals in charge of organizations can try to control the relationships between members, manipulating feedback loops or modifying reporting lines, but the person, or people, instituting those changes will be affected by the changes as well. Their relationship to others in the group will be dramatically altered when people see these changes being instituted. Unintended consequences are very likely to result, since it is impossible to remove oneself from a system one is a part of. Going outside the system to create change won't work, either, since the change agent could be viewed with suspicion and distrust by those who now see these people as outsiders.

The best we can hope for, it seems to me, is an increased understanding and awareness of the natural dynamics of groups so that all of us, as individuals, better understand what is going on around, and among, us. This, essentially, raises our IQ for group behavior and the more of us who can do this, the more likely it is that the group's behavior will eventually change. Any attempt to apply the insights of self-organization to groups of unsuspecting people is apt to backfire, and the reason is quite simple: the collective is smarter than you are.

We have seen this fact demonstrated in dramatic fashion in recent years. Our political and economic systems, especially, seem to function according to their own rules of behavior, rules that nobody has any control over. This much is clear, and yet people continue to believe that somebody, somewhere, must be in charge of all this and that person or small group is simply refusing to regulate or manage the system that seems to be behaving in ways that none of us understand. It's simply not true. No one person, or even a small group of people, is in charge of our social, economic and political systems. Some people understand how these systems work better than others do, but nobody has the control of any levers that can be pulled or buttons that might be pushed that will control everything that these systems do. We are, all of us, individuals in a great collective of humanity that is growing and evolving in ways that will lead us to a new type of existence as a species — a new existence that is wholly unknown to any one of us.

Indeed, this has already happened. Consider who we were a mere 6,000 years ago — a blink of the eye in the history of humanity, and a mere flutter of the eyelash in geologic time. A few thousand years ago, small groups of people were just beginning to develop written languages, there were no countries, no political or economic systems, no means of mass communication. There were tribes and battles between them (that much hasn't changed) but little we would now identify as literature, theater, politics, economics, much less science or medicine, existed — and yet the humans who lived at this time were virtually identical to us in a biological sense.

One difference between us and our ancestors of 6,000 years ago is that they only lived 35–40 years on average. This is one dramatic piece of evidence about how our collective intelligence has resulted in improvements for all of us: the average human lifespan has doubled almost everywhere in the world. While this is, sadly, not yet true for all humanity, we have the means to make it true for everybody. The medical and technological advances that have made this doubled lifespan possible are the result of self-organization of our species. No one person, not even a small group of people, has been in charge of this continuously evolving development of methods for improving our collective lifestyle.

There have been visionaries, to be sure, people who proposed new concepts about the sources of diseases, or the rights of people, or who found a new way to print multiple copies of texts, but the rest of our species was smart enough to follow these visionaries and flock toward the promise of the future their insights provided. Nobody forced us to do this; it is self-organization that has created human society. Without our societal inventions and creations, our species would be composed of entirely different creatures.

None of this change came about without turmoil and struggle and suffering, but the thrust of the overall effort has been largely, although not totally, positive. We are now beginning to see that our drive toward advanced technologies has resulted in damage to the environment we live in. Changes are being proposed to deal with this damage and we still have a chance to do something about it. Addressing climate change doesn't mean that all technological activity should halt (how could we stop what is really our biological imperative, anyway?) but it does mean that we will have to use our collective smarts in new ways as we continue to self-organize.

In my view, the best way to weather the storm of self-organization is to *trust*. Rather than trying to control uncontrollable systems that we may be a part of, I think it is better to consider the possibility that all this turmoil and strife in the world is evidence that a better life or existence *for us all* is taking shape and that a benevolent force is behind it all. This can be difficult enough when the self-organizing system we are caught up in is a family or a church, but what if it seems to be our nation — or the whole world — that is undergoing a possible self-organization? How do we even know that it *is* self-organization that is occurring and not some apocalyptic event? When cataclysmic change is occurring all around us, when mighty skyscrapers crash to the ground and deadly diseases ravage the population, is our world self-organizing? Or is our world coming to an end?

Or is this just two different ways of saying the same thing?

Just as those early single-celled individuals would have told us, and just as every caterpillar knows who emerges from her cocoon to find she has somehow become a butterfly, there is no way for us, as individuals, to tell the difference between the end of a familiar way of life and the beginning of a whole new, wonderful world. It is at times like this, times when one attractor shifts to another, times of bifurcation, that our knowledge and understanding of self-organization can serve us well, and provide the comforting knowledge that all of this — *all* of it — is simply a sign that we are growing, and that Creation continues unabated.

Endnotes

1. Walt Whitman, "Sea-Drift: On the Beach at Night Alone," in *Leaves of Grass*, 1882 Edition: https://en.wikisource.org/wiki/Leaves_of_Grass_(1882). This poem is in the public domain.

2. A good book that covers the BZ reaction as well as other chemical oscillators is: Irving R. Epstein and John A. Pojman, *An Introduction to Nonlinear Chemical Dynamics: Oscillations, Waves, Pattern and Chaos*, Oxford University Press, Oxford (1998).

3. Stanley L. Miller, "A Production of Amino Acids under Possible Primitive Earth Conditions," *Science*, Vol. 117, Issue 3046, pp. 528–529 (1953).

4. Engineers have used our understanding of flocking behavior in animals to design and control engineered systems. See, for example, Roland Bouffanais, *Design and Control of Swarm Dynamics*, Springer Briefs in Complexity, Springer, Singapore (2016).

5. Dirk Helbing *et al.*, "Improving Pilgrim Safety During the Hajj: An Analytical and Operational Research Approach," *Interfaces*, Vol. 46, No. 1, pp. 74–90 (2016).

6. For a more thorough account, see for example: Alistair E. McGrath, *Science and Religion: An Introduction*, 3rd Edition, Wiley-Blackwell Publishing, New Jersey (2020).

7. Pantheism literally means "all is God", so every material thing in the universe is divine. Panentheism, on the other hand, says that "all is God, but the divine extends beyond space and time." See for example: John Haught, *A John Haught Reader: Essential Writings on Science and Faith*, Wipf and Stock, New York (2018).

8. The most widely cited work is that of Sydney Fox, who carried out experiments to extend the Miller-Urey work. See:

   (a) S.W. Fox, "Origin of the Cell: Experiments and Premises," *Naturwissenschaften*, Vol. 60, pp. 359–368 (1973).

   (b) S.W. Fox, K. Harada, G. Krampitz and G. Mueller, "Chemical Origins of Cells," *Chemical and Engineering News*, pp. 80–94 (1970).

   (c) S.W. Fox, "The Protein Theory of the Origin of Life," *American Biology Teacher*, Vol. 36, pp. 161–172 (1974).

9. (a) G. Gerisch, "Cell Aggregation and Differentiation in Dictyostelium," *Current Topics in Developmental Biology*, Vol. 3, pp. 157–197 (1968).

   (b) J.J. Tyson and J.D. Murray, "Cyclic AMP Waves During Aggregation of Dictyostelium Amoebae," *Development*, Vol. 106, pp. 421–426 (1989).

10. A video of evolving waves in the BZ reaction can be viewed on YouTube: https://www.youtube.com/watch?v=jRQAndvF4sM.

11. His most widely cited paper is: Otto E. Rossler, "An Equation for Continuous Chaos," *Physics Letters*, Vol. 57A, No. 5, pp. 397–398 (1976).

12. Any textbook on physical chemistry will have a good explanation of the laws of thermodynamics. A classic is: Walter J. Moore, *Physical Chemistry*, 4th Edition, Prentice-Hall, New Jersey (1972).

13. For a detailed summary of his ideas, see: Gregoire Nicolis and Ilya Prigogine, *Self-Organization in Nonequilibrium Systems: From Dissipative Structures to Order through Fluctuations*, John Wiley & Sons, New York (1977).

14. Stuart Kauffmann, *The Origins of Order: Self-Organization and Selection in Evolution*, Oxford University Press, Oxford (1993).

15. Pierre Teilhard de Chardin, *The Phenomenon of Man*, Harper and Row, New York (1959).

16. Margaret J. Wheatley, *Leadership and the New Science: Learning about Organization from an Orderly Universe*, Bernett-Koehler, San Francisco (1992).

Insight 4   Cycles & Rhythms

Whirling around the Still Point of Ecstasy
I spun like the wheel of Heaven.

Rumi[1]

Like most people who find their way to science, I've never had any trouble coming up with questions about life or the world around me. I soon found, however, that it wasn't enough just to say I was curious and wanted to know more — it wasn't enough to say that I just had this "feeling" that studying these things would lead me to the answer to my Big Question. I'd had this question since childhood and it was, simply, "Who am I and why am I here?"

For one thing, I quickly learned that nobody cared much about my Big Question. If I wanted any help in finding the answer, any money to pay my bills, any colleagues to talk to about it, I needed to reduce my question to smaller, more manageable pieces. I couldn't just ask, "Who am I and why am I here?" Instead, I tried to ask, "What makes this collection of molecules that is *me* alive?" and followed that up with, "What makes *any* collection of molecules alive?" I wanted to know how a bunch of molecules such as my own could get themselves so organized as to even frame such a question, but that seemed beyond my grasp, so I settled for, "Why does this system of molecules in the BZ reaction *act* like it's alive?"

In addition to slicing my sweeping questions into smaller questions that could be answered with the tools of science, I needed to rephrase them to be more precise. Instead of, "Why does the BZ reaction act like it's alive?" I needed to ask, "What dynamical features of the chemical kinetic mechanism lead to the existence of temporal organization in the form of oscillatory behavior?" I needed to speak the language of science.

Beyond limiting the scope of my inquiry, then rephrasing my questions more precisely by using accurate technical language, I also found that I needed to prove the uniqueness of my approach to finding answers to these increasingly precise questions. Even though I had been lucky enough to finally land a tenure-track faculty position after a multi-year search, I still needed funding to pursue my research. I quickly learned that to convince funding agencies to share some of their highly sought-after research dollars, I had to persuade the reviewers that what I wanted to study had, first, never before been investigated; second, was so important that they just had to provide financing to study it; and third, that I was the best person for the job. I got so involved in both the quest to find resources to support my work and the myriad problems that arose in running a lab and hiring research assistants that, within a few years, I had almost forgotten *why* I was doing all this.

This part of my story is a nearly universal feature of the career path for most scientists I know. I didn't realize, when I first considered becoming a scientist, that so much of the scientific effort is really about marketing. Marketing your ideas, selling the importance of your work, persuading journal editors or program officers or deans to provide resources such as money, space, even a publication platform. Securing these scarce resources is the day-to-day concern of all scientists, and I was no different.

In occasional moments of clarity, I would remember what had once motivated me, and wonder if, somewhere beneath my to-do lists, that girl who had gazed up at the Milky Way, contemplating her Big Question, still existed. In those flashes of remembering how I had gotten where I was, I would wonder if I was actually doing what I was meant to do — or had I, somehow, gotten horribly off track?

I still had a strong sense of being pulled along by a force that I didn't have much control over — and this force was compelling me to learn, and learn, and learn some more. Furthermore, once I had learned something well enough, I felt I needed to tell the story about it in as many venues as possible — in research papers, in talks and in grant proposals. I didn't talk about my sense of being drawn along by an unseen force with any of my colleagues. I had known as early as grade school that a lot of other kids didn't share my compulsion to learn all I could about the universe. Even though I had finally found colleagues who exhibited at least some of that same desire, I still wondered if I was the only one who felt like she was under the influence of an uncontrollable force. I knew in my bones that this force — an attractor, as I eventually came to think of it — was pulling me deeper into an ever-expanding world of intricate and detailed knowledge. It was to be many years before I understood how those breakthrough moments, like the ones that hit me as a child gazing at the Milky Way, were rare but incredibly valuable, glimpses of the force at work as it drew me deeper and deeper into study and gave me fuel for the quest.

Rhythms in an Obscure Root Vegetable

As I gained more experience in science, I learned not only to make my questions more precise but also how to determine which of these questions were truly interesting and which were not worth my time and energy — not worth anybody's time and energy, actually. I gradually turned the focus of my work to one particular chemical reaction that occurs in the root of the horseradish plant. Horseradish root might not seem like it would be a source of "truly interesting" scientific questions, and, perhaps, might be worthy of one of the late Senator Proxmire's infamous Golden Fleece Awards.[2] In the 1970s and 80s, the US Senator from Wisconsin handed out these "awards" to those studies he found wasteful of government spending. The recipients were almost always "obscure" scientific projects, and I'm sure they often seemed, to an outsider, to fall into the category of "not worth anybody's time and energy". This mocking approach to criticizing the way science is done has, sadly, continued even after Proxmire left the Senate and been taken up by other short-sighted politicians.

So, although I sometimes worried that my work might attract unwanted attention from the likes of Senator Proxmire, I really didn't think it qualified for one of his awards. The reasons the humble horseradish plant fascinated me were really quite simple: first, the chemical reaction in horseradish root I was interested in had already been studied a lot, so there was a foundation of knowledge upon which my own work could build. Second, and more important for my purposes, it exhibited chemical oscillations[3] (see Figure 8) similar to those red and blue oscillations I had first observed in the BZ reaction.

But — and this was the key point — it was not the BZ reaction. If the insights people were achieving about the inorganic BZ reaction were also occurring in a living plant, studying and comparing the horseradish system with the BZ might

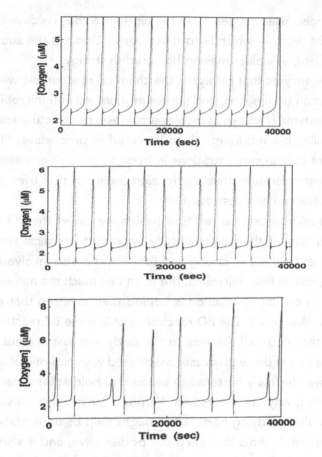

Figure 8: Examples of oscillations in the oxygen level in a computer model of the PO reaction. The top two panels are periodic oscillations; the bottom panel is chaos.

reveal something important about the nature of life. Without the horseradish example, or something like it, we would never know if the behavior of the red-blue BZ reaction was just a quirk due to its specific chemical nature, or if it had more fundamental importance for our understanding of the role that cycles and rhythms play in living things — like ourselves.

Other Oscillations in Living Things

The specific chemical reaction in horseradish we were studying is called the peroxidase-oxidase reaction because it's controlled by an enzyme that normally uses hydrogen peroxide to do its work (that's where the "peroxidase" part of the name comes from) but in this one little part of the inner workings of the horseradish it also uses oxygen from the air (hence the "oxidase" part of the name) to do its work — which is to make lignin. Lignin is the substance that makes wood hard, as well as making horseradish stringy.

The type of enzyme that catalyzes the chemical reaction we were studying is found in human bodies, too, and serves as a natural anti-microbial agent. In humans, this enzyme is known as myeloperoxidase. Its structure and molecular action are similar, but not identical, to horseradish peroxidase. Myeloperoxidase is present in high concentrations in body fluids such as saliva and tears, where its presence ensures that any foreign organism that tries to enter the body will be attacked and destroyed.

So the peroxidase-oxidase reaction (which we called the PO reaction) in horseradish is one of the fundamentally important chemical reactions that make trees what they are — strong and tall — and is also involved in the process that protects us from microbes. But it isn't so much the nature of trees or the fact that an oscillating reaction is behind their strength that encouraged me to keep working on it. The PO reaction, just like the BZ reaction, is similar to reactions[4] that occur all the time in the body, not just in trees and horseradish and not just in those processes associated with anti-microbial activities.

For example, the lowly horseradish seemed to hold at least part of the key to understanding why our hearts beat,[5] among other things. It was an exciting possibility that studying horseradish might help us understand what was behind the crucial rhythms that keep our bodies alive, and it seemed to aim my work back toward my Big Question — even though, by the time I was narrowing my focus to the PO reaction, I had nearly forgotten that I ever had such a lofty question. Fortunately, the bifurcation I was to encounter several years later in my own life reminded me quite forcefully *why* I was doing science in the first place — and that it had nothing, and everything, to do with horseradish.

Focusing my attention on this obscure chemical process in a plant that few people cared about yielded a gold mine of insights. Through it, and through comparing what we learned about it to what others were learning about the BZ reaction, I was led to a deepening appreciation and understanding of important concepts like self-organization, the attractor, bifurcations, and the fundamental stability of cyclic or oscillatory processes in living things. Through this work, I began to see that study of the lowly horseradish could provide profound lessons about the most basic questions we humans have about life. In fact, if it weren't for horseradish, I never would have found my way back to my Big Question.

As I got deeper into the study of cyclic behavior in horseradish root, I began to see just how often living things exhibit their true dynamic nature in a cyclic way: cells divide every couple of hours, hormone levels in our bloodstream spike four or five times a day,[6] our bodies go through daily and monthly cycles and plants grow and bloom on a seasonal schedule that is roughly the same year in and year out. In the winter, plants may appear to be dead, but we know they are only in the resting phase of their cycle, waiting for the annual period of growth to repeat as it does every spring. So, as I studied the cycles occurring in the horseradish root, I began to realize that by studying this one plant I might be able to find out a bit about where all those other cycles in living things came from.

It is not always easy to tell life from death, but those underlying cycles can help us do this. Even the old walnut tree outside my window, which stands quite still and unmoving unless the wind stirs its branches, might be mistaken for being dead in the winter. However, it is constantly changing — taking in water, carbon dioxide and sunlight, converting these raw materials into cellulose and other cellular components, sending out new shoots of growth, etc. These changes, although they occur on a scale too small to be seen and too slow to be noticed, tell me that the tree is alive, and caught up in cycles of growth that match the seasons.

Not all things that behave in a cyclic manner are alive, though. Consider, for example, the cyclic red and blue flashes in the BZ reaction, or even the cycles of the moon around the earth — both oscillatory phenomena of non-living things. The more I considered the question, "What makes something alive?" the more I realized that the reverse statement does seem to be true: living systems always move forward through time in a cyclic manner, not a linear fashion.

Cyclic dynamics are, in fact, a fundamental characteristic of life at all levels. There is the cell division cycle I mentioned earlier and, at a slightly higher level of organization, circadian rhythms that govern our sleep/wake cycle, oscillations in voltage in cardiac pacemaker cells that produce the heartbeat, and others. Similar oscillations in groups of networked neurons produce thought and memory (exactly how they do this is still the subject of intense scientific study). There are also whole-body biorhythms in hormone levels, blood sugar levels, and many other measures; and, beyond the level of a single organism, there are oscillations that arise in groups or whole populations of organisms that are even more astounding.

The 17-Year Brood

I learned a dramatic lesson about cyclic dynamics one summer in Indiana that started off like all the others. We had lots of heat and humidity that year, lots of rain, and by early July, the air was filled with a sound I had come to associate with the typical Midwestern summer: a strong thrumming noise that originated from the upper tree branches and got louder and louder as the month went on. The sound was a sign that a rather ugly creature, the Cicada, had started to make its annual appearance. Each year in late June, hordes of these insects crawl from their underground winter homes near tree roots and continue on up the trunks to the topmost branches where the males begin to "sing" (if you can call it that), trying to attract a mate.

While I had come to associate summer in Indiana with these loud, pulsing chirps of the Cicada, that year the noise was distinctly different. For one thing, it was much louder than in all the previous years, and the thrumming sounded different, like some kind of alien spaceship. It would increase throughout the afternoon until it got so loud we couldn't carry on a conversation outside. The Cicada, which look a bit like giant houseflies, began falling by the handfuls from the branches and tree trunks, and it quickly became clear that there were a lot more of them than in previous years. Not only couldn't we speak to each other outside, a simple walk from the house to the car or bus forced us to step on a crunchy carpet of fallen bug bodies that formed beneath every tree.

As I soon learned, that year corresponded to the peak in the collective life cycle of the 17-year periodic Cicada, an insect that spends most of its life underground as a dormant nymph, emerging only after having been buried for 17 years. In one summer season, they mature rapidly, lay eggs, then die. The population of nymphs, known as a brood, had been placed there by their mothers as fertilized eggs 17 years before. That year, they went through an amazing transformation, rising up from the ground *en masse*, sprouting wings, flying around and mating in a frenzy of activity before laying their own eggs and finally dying. If we had unearthed the Cicada nymphs earlier in the year, we might have thought they were dead, too, but the frenzy of life activity showed us that they most definitely were not. They were vividly and loudly alive, at least for five or six weeks in the summer every 17 years.

Heart, Brain, and Calcium

I was intensely interested not only in the cycles occurring in our horseradish plants, but in the plethora of cyclic behavior I started to notice around me, including the dramatic Cicada explosion in our trees. At the time I was deep into the study of horseradish, some of my colleagues were discovering other chemical processes occurring in plants and animals that were similar to our PO reaction. One of these processes transported electrolytes across heart cell membranes in a cyclic manner. A recent discovery had revealed that this cyclic electrolyte transport caused pulses of electricity to be produced by the pacemaker cells of the heart. These pulses, in turn, caused the heart muscle to squeeze rhythmically, one beat for each pulse of electricity. This was a very exciting result, and seemed to indicate that an oscillating chemical process might be the cause of our heartbeat and hold the key to understanding heart arrhythmias.[7]

If I hadn't already been intensely interested in chemical oscillations, the new data about cardiac pacemaker cells were more than I needed to delve in further. It wasn't just these pulsating heart cells that intrigued me — oscillations were cropping up everywhere I turned. Some papers I read at the time reported oscillations in cell types other than heart cells — pancreas cells, I learned, produce insulin in pulsating, rhythmic squirts. These insulin oscillations, it turned out, were regulated by oscillations in calcium concentration within the pancreas cells. Another set of experiments had shown that insulin levels oscillate in the bloodstream throughout the day — as do many other regulatory chemicals. So, yet again, I was finding that another fundamental biorhythm seemed to be controlled by an underlying chemical oscillation. Other discoveries soon convinced me that cellular calcium levels in a whole host of cell types oscillated throughout the day, producing an internal symphony pervading every cell in our bodies. Liver, muscle, egg cells, and many other types of cells were all found to be pulsating packets of calcium.[8]

This result about the pulsations of calcium in our cells was especially exciting to me — I knew that calcium in the cell interior serves as a "chemical switch" in the body. Changes in its concentration have long been known to be the trigger which turns off and on many important cellular processes, including those that translate genetic information in the chromosomes into biological characteristics — blue eyes or brown, fair skin or freckles, even a genetic predisposition to disease. These discoveries about calcium oscillations, which were due to much the same process as the one that made our horseradish root oscillate, showed me that the chemical process my students and I were studying had widespread effects on many of the basic functions of life.

Predator-Prey Dynamics

The spruce budworm example from an earlier chapter shows how unusual behavior can arise in systems where a predator and its prey are interacting. In that example, the budworm population can exhibit what is known as bistability: two stable states that arise when the underlying attractor for the system bifurcates once a critical carrying capacity is reached.

Stable oscillations are another way predator-prey systems behave. This is now a well-known phenomenon and guides the work of wildlife biologists who must manage populations where one species is intent on eating another one.

The dynamics of predator-prey systems were not always so well-understood, however. In the mid-1920s, Alfred Lotka, a US mathematician and physical chemist, and Vito Volterra, an Italian mathematician, independently introduced a set of equations that have since become known as the Lotka-Volterra model.[9] These equations describe the interactions of a system of predator and prey and explain why these types of systems are most stable when the predator and prey populations oscillate out of phase with one another.

This was the first time such a thing had been proposed but it had wide-reaching consequences. In addition to studying populations such as a mix of rabbit and fox, Lotka considered epidemics and showed that these could be modeled in similar ways. A virus is, in some ways, a predator seeking out its prey, a host that it can infect.

Vito Volterra was simultaneously studying and trying to explain the unusual patterns in populations of fish species in the Adriatic Sea.[10] An Italian biologist had noted that more of a certain type of predator fish were caught in the Adriatic Sea during World War I when many fishermen were at war and unable to work. This puzzle intrigued Volterra who developed a model to explain it: the predator population surged to higher levels shortly after the time the population of prey it fed on had increased. It was a natural reaction to more food.

The problem was the hungry predator fish quickly depleted their own food supply, reducing the predator population and allowing the prey to reproduce and build up again. It was a natural oscillation process that reflected these two species essentially "taking turns" as they populated this part of the ocean. This oscillation was not only natural, it was stable.

The equations developed independently by Lotka and Volterra have since been applied to a number of predator-prey systems and have been used to explain puzzling phenomena such as the oscillations observed in the number of pelts from hare (prey) and lynx (predator) gathered by the Hudson Bay fur-trading company starting in the late 1800s, as shown in Figure 9.

Long before we ever did the experiments which led to our understanding that a stabilizing attractor governed the oscillations in the PO reaction, we had a sense that this chemical reaction would reveal an underlying order, since the test tube was filled with molecules which behaved, in a sense, like a collection of foxes and rabbits trapped together in a barnyard.

One of the components of the PO reaction are molecules that behave just like a predator, say a fox, and others that tend to behave like prey, rabbits. The fox-like molecules "devour" the rabbit-like ones who, in turn, (behaving just like rabbits) will produce multiple copies of themselves and increase without bound if no fox-like molecules are present.

The rabbit-like molecules are produced by a type of chemical reaction known as autocatalysis. An autocatalytic reaction is one in which the product catalyzes its own formation, so one molecule will produce more than one. Mathematically, this follows the same dynamic laws as an adult rabbit producing more than one baby rabbit. Notice that if there were no fox-like molecules to "consume" those rabbit-like ones, we would be dealing with an explosion — not an oscillation.

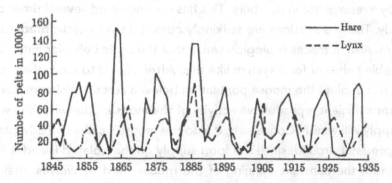

Figure 9: Historical data gathered from a fur-trading company showing natural predator-prey oscillations.

An interesting real-life example of a predator-prey system is found on a remote island in northern Michigan, Isle Royale.[11] The Lotka-Volterra equations have been used to understand this isolated island system, which is the home of a substantial population of wolves and nearly 2,000 moose. Ecologists who study the wolf and moose inhabitants of Isle Royale have found that the populations of these two key species are not constant; at certain times, the wolves experience a surge in population growth, while at other times, the moose reproduce rapidly. The surges in populations of the two species alternate: an increase in the number of moose is inevitably followed by an increase in wolves; the rise in the number of predators (wolves) then causes the prey (moose) population to plummet. The same is true of the wolf population, but at a time delay: an increase in the number of wolves will deplete the moose population, decreasing the wolves' food supply and, thus, eventually decreasing the wolf population. The populations of wolf and moose follow alternating cyclic patterns in which prey increases as predator decreases and vice versa.

Ecologists who first observed these alternating surges in population were alarmed each time the wolf population began to drop, sometimes falling to very low numbers: viruses and genetic problems due to extensive inbreeding were suggested as possible reasons each time a decline began. When the wolf population became dangerously low, the ecologists feared that the wolves would go entirely extinct on this island, an end result nobody wanted, as it would lead to certain ecological disaster. After all, without wolves, the moose would quickly outstrip their limited food supply.

Unwilling to interfere with the object of their study, however, the ecologists held back, doing nothing but counting individuals, both predator and prey. Each time the plunge in the number of wolves occurred it was, inevitably, followed by a resurgence in numbers. This has now occurred several times during the study. The observations are strikingly consistent with cyclic predator-prey dynamics. Also, it makes ecological sense that the cyclic behavior might be the most stable behavior for a system like this. After all, it is to the wolves' advantage to not deplete the moose population below a certain minimum level necessary for sufficient reproductive activity of the moose; otherwise, the wolves' food supply will vanish. Conversely, it is to the moose's collective advantage to have a predator around, since the food supply on the isolated island is limited and without the constant thinning of the moose herd by wolves, their food supply would soon be depleted, starvation would set in and the entire herd could die.

These two trends have opposite effects on the dynamics of a predator-prey system. What is, perhaps, most surprising is that the populations of predator and prey do not, even eventually, settle down to a constant, optimum value after several alternating surges, but, rather end up in an endless cyclic motion — but one which is, nevertheless, stable. The paradigm of homeostasis, in which a living system finds a balance, could lead to an incorrect understanding of this type of system; stability in a predator-prey system is *only* achieved when the populations of predator and prey are both cycling — each one growing abundant, then growing scarce, only to repeat the pattern over and over. The danger of not understanding this type of dynamics and its inherent stability is the all-too-human urge to do something about the plummeting populations which seem to be threatening extinction. The urge is laudable but the cure may not be warranted by the symptom.

Round and Round through Life We Go

As I started to pursue my own research program, for the first time independently of the professors I had studied under, my intense interest in the obscure process in horseradish soon widened into an investigation of biorhythms of all types: calcium oscillations, cardiac pacemaker dynamics, even those oscillations in voltage that occur in our brains when groups of neurons fire in synchrony. After several years of working with these scientific techniques in my own lab, developing detailed theories of all sorts of biological oscillations, I began to wonder if the cycles I sensed in my inner life — my emotional and spiritual life — could also be described with similar mathematics.

These cycles, which we all sense, occur whenever we find ourselves in a recurring emotional or spiritual state, learning (or not learning) the same lesson, again and again. *Haven't I been here before?* I would hear myself say. *Didn't I learn this painful lesson already? Why do I have to go through all this again?* It was only after I had gone through a life-shattering bifurcation event that I began to realize the cycles had always been there in my own life — and that this inner pulse, far from being a sign of inadequacy or my own inability to stop going in circles, was evidence of a deep organizing force at work in my life — indeed, in all our lives.

A lot of my own sense of inner cycles came from a vicious circle I kept falling into. It was a common one for women in my generation, characterized by its two extreme polarities: my work at one pole, my family at the other.

I was constantly pulled in two directions and felt perpetually guilty. When I was working, I worried I should be with the kids. When I was with the kids or taking care of things at home, I felt I was neglecting my work. It often seemed like I was caught in one of the oscillations I'd been studying in the lab, but I never had time to give this strange notion more than a passing thought.

I tried to spend as much time as I could on my research — but all I could really manage were small moments I found between teaching, caring for a home and raising two active kids. Juggling everything I needed to do was difficult, but I persisted and a good deal of my motivation to find ways to keep my research going was a strong compulsion to understand the systems I was investigating. I really wanted to understand what made them do the amazing things they did — things like bursting into spontaneous oscillations or producing elaborate, beautiful patterns. I thought I could understand why they did those amazing things if I just worked hard enough and long enough.

Somewhere under this almost mindless drive to get deeply into the workings of my chosen system of study, an obscure chemical process in a lowly root vegetable, a mindlessness driven by little sleep and long work hours both at the lab and at home, was a vague memory of my Big Question from childhood that had sparked my journey into science. I didn't consciously remember it anymore, but it still influenced my choices about how I spent my time. Were it not for the journal I'd kept since high school, throughout college and the first part of graduate school (before my first child was born), I doubt I would have ever known that the person I used to be before diapers and grading papers and writing grant proposals had looked out at the stars and wondered who she was, and why she was here.

As my husband and kids and I established a life together in Indiana, we fell into a routine, not unlike most families: daily events that happened at roughly the same time each day, weekly and monthly schedules, even annual cycles tied to the school year. In a way it was comforting since I knew where I was supposed to be at any given time and what I was supposed to be doing. If it was 7AM, I should be sending one child out to the school bus and buckling another into his car seat for a trip to the daycare center. If it was Tuesday, I should be on my way to deliver a lecture to several hundred chemistry students, but if it was Friday, I should be preparing for a faculty meeting.

The annual cycles were also quite predictable: if it was August, you would find me writing a syllabus for whatever course I was teaching, but if it was December,

I was grading final exams. The predictability of it all meant that I didn't have to think much about what I was supposed to do at any particular time.

After some years of the same routine, I realized I was approaching the beginning of a semester with a sense of dread. It was no longer comforting. I had a sense of rewinding the tape, ready to hit the start button on the same sequence of lectures I had delivered several dozen times already.

Even though predictability and reliability are comforting aspects of the cycles of life, some variation is desirable to keep us interested. Remember the movie *Groundhog Day?* In this film, Bill Murray finds that he is doomed to repeat February 2nd over and over, each day exactly like the one before, until he learns an important life lesson that will allow him to break out of the cyclic, predictable pattern. A day like his becomes, after endless repetitions, no longer comforting in its predictability but deadly in its sameness.

While it is true that our lives, both physiologically and spiritually, are inherently cyclic, characterized by pulsations and rhythms at all levels of our existence, it was to be many years before I could see how the monotonous sameness of a repetitive cycle can, at times, be comforting in a deep, fundamental way — the way the seasons, for example, are not boring to us but, rather, provide evidence of a stable rhythmic foundation upon which a life full of creativity and surprise can thrive.

Trusting in the Rhythms

The transition from winter to spring has gone on for millennia with great predictability, convincing us that we live in a deeply stable, although clearly cyclic world, despite the daily or weekly irregularities we call "weather". It's interesting to consider whether our own inner cycles are as stable as those of, say, the seasons, for instance, or the motion of the planets around the sun. Both of these cyclic behaviors are the very essence of stability and predictability, yet neither is static. I find comfort in the notion that even the darkest, bleakest winter will always give way to a warm, green spring and that Venus will always appear in the sky at a position even the earliest astronomers had learned to predict thousands of years ago. Could I believe as much about my own inner cycles?

At a deep level, all of our lives are characterized by rhythms, and it was helpful, from time to time, to return to this fact and remind myself of the lesson of stability that these fundamental rhythms provide. Although it's getting a bit ahead of the story, I eventually discovered that contact with this inner rhythm, rock-solid in its stability, was easily and immediately accessible at any moment, in any place, by merely focusing on the rhythm that was closest at hand: my breath.

As I unconsciously inhale and exhale, I am actually participating in a fundamental spiritual practice: showing trust. For most of my life, of course, I never thought of mere breathing as a spiritual practice, but as my life began to rearrange itself after the bifurcation event in the driveway, I found my way to people and situations where I was taught how to turn my awareness inward, and the first instruction was always the same: focus on your breath.

When we breathe, we trust that if we exhale, let go of the air we had just moments before so eagerly inhaled, another cycle of breath will bring fresh oxygen into our lungs. We do this unconsciously, but in a trusting way. We trust that our breath will continue whether we consciously decide to inhale or not. We also trust that our bodies will exhale automatically, removing the waste products produced by our cells as they use that oxygen to give us life energy, waste that would surely kill us if it were not quickly removed. Most of us don't think about the fact that we trust our bodies every moment to serve us this way, but it is true: we trust that this inherent rhythm of life will continue whether we remember to breathe or not.

The level of trust we must achieve every time we exhale is not unlike the level of trust we might imagine a tree must have as autumn approaches and it must pull the life force in from the ends of its branches, letting the leaves drop. As the leaves fall, the tree retracts vital life energy that had flowed into those leaves all spring and summer, drawing it deep down inside itself to store it away in its inner core and roots for the winter. Just as the tree must let go of its leaves, trusting that the cycle will go around later and the leaves will come back, we must trust that if we exhale, the cycle will go around and new, fresh air will come back into our lungs.

The sense of underlying cycles is a common element in many religions and spiritual philosophies. A central feature of Buddhist thought, for example, is the endless cycle of birth, death and reincarnation known as samsara.[12] The idea of a reincarnation cycle did not originate with Buddhism, though. It was a widespread belief in India before Buddhism came into existence and is a central feature of the Hindu tradition, as well as other religions.

Even in the Western Judeo-Christian tradition, the great cycle of life is celebrated, as this well-known passage from Ecclesiastes (3:17) says so poetically: "To everything there is a season, and a time for every purpose under heaven." The belief in a great cycle of life continued to be widespread, even in the West, until it began to strike some as too pagan and, therefore, supposedly incompatible with Christian belief. In the last several decades, however, we have seen a resurgence of interest in these ideas most obvious in the increasing numbers of Western peoples involved in Eastern spiritual practices such as yoga and meditation.

Meditation can be quite difficult for those of us who have a hard time quieting the mind, and some meditation traditions deal with this natural human tendency by counseling the meditator to focus on the breath. My own introduction to meditation started this way, and while focusing on the breath certainly worked to quiet my mind a bit, I soon discovered that it was also a wonderful device for bringing me into immediate and direct contact with the inherent cyclic nature of my life.

I would begin my meditation by imagining my body as a tree: my spine as the trunk, my legs and feet as the roots and my neck, head and arms as the branches. With each exhalation, I would visualize my tree body letting go of leaves that sprouted from my arms and shoulders, pulling life energy into my trunk and down into my roots. Then, when the inhalation came naturally some moments later, I visualized that same life force rising up from my root and into my spine where it could flow into my arms and hands, bursting forth as new leaves.

The lesson I learned each time I participated in this simple mediation was two-fold: I could trust my breath to know what to do from moment to moment, with no mental effort on my part; and, secondly, by meditating on the tree's annual cycle of leafing out, then dropping leaves, I learned to *let go*.

Learning to let go might seem easy enough, but if you are afraid of what might come up if you do (as I was), you will probably hang on as tightly as you can. This meditation was helpful to me because it showed, in an almost experimental fashion, that I was wrong in thinking that I needed to be in control of every aspect of my life. In a strange way, focusing on my breath showed me that I didn't really have to think about breathing — it just happened. And, by extension, maybe I didn't really have to think about, or control, every minute detail of my life. This was difficult to do, when the inner process that was set in motion that day at my mailbox began to tug at the threads I had woven together into a life, and I felt I could do nothing but stand back and watch as everything unraveled. Under these conditions, it can be very, very difficult to let go.

Stable does not mean Static

Dynamic behavior, particularly *cyclic* dynamics such as the pulsating calcium concentrations in pancreas cells, our heartbeat, or the 17-year cycle of the Cicada, is a characteristic feature of life. The opposite of a dynamic system is a static system — that is, one which never changes — and we would rightly relegate static systems to the realm of the inert or dead. However, our geologist colleagues would warn us to be careful about choosing examples of supposedly static things. After all, what seems more static and unchanging than a rock or mountain? And yet, on a geological time scale, those apparently unchanging rocks and mountains are actually constantly evolving. The earth is very dynamic indeed.

I will leave it to others to speculate about whether this means the earth is alive — but it should give us pause to realize just how rare the truly static and unchanging really are. Even atoms and subatomic particles are constantly moving and vibrating. The universe, all of it, is in constant motion.

Since dynamic behavior is a lot more interesting than stasis, it is surprising how often we fight against and reject it. When faced with obvious signs of change, we yearn for things to stop. When our life or the world around us is rapidly evolving we do, in fact, feel much more alive than usual, but most of us wish, during these times, for more obvious signs of stability. If our business or organization is changing so rapidly or erratically we don't know what will happen next, or our life is so unsettled that we never know from one day to the next how we will feel or what we will do — well, *this* kind of dynamic behavior is something we would rather do without. At times of great change, we think we yearn for equilibrium — we want everything to just stop changing so we can get our bearings again. Things would be so much more stable that way, we think.

Many of us have the notion that a sign of stability is equilibrium. At equilibrium, a system has reached a steady operating condition, never changing after reaching the equilibrium state. We somehow believe that equilibrium[13] should be our goal. After all, if things would just stop changing so fast, life would seem so much more stable. But true stasis would mean we're dead, so at the same time we yearn for equilibrium, we know we don't *really* want it. We seem to want equilibrium, but one that is not dead, not static, still full of life, somehow.

I believe the problem here is that we (as individuals *and* organizations) often confuse *equilibrium* (which is quiescent, static, unchanging behavior) with *stability*, but the two are completely different ideas. Equilibrium, i.e., a static unchanging pattern, is stable, but *dynamic patterns* can also be stable. It is *stability* we yearn for, not *stasis*. Cyclic behavior, such as those oscillations I observed when studying the chemical reaction in horseradish, or oscillations in predator and prey populations, can be completely stable.

Stable cyclic motion will resume even if the system is temporarily buffeted by outside forces that throw it off its cycle for a while. This is because the oscillating system is governed by an attractor that damps out any fluctuations and pulls the system back to its stable, oscillating state. This type of attractor is known as a limit cycle attractor (see Figure 10). It has this name because all nearby trajectories will, given enough time (or, as mathematicians say, in the limit of infinite time), make their way to that attractor and stay there.

To illustrate through a gross exaggeration the disaster which could ensue if we were to look on every instance of cyclic behavior as undesirable, something to be "fixed", consider the seasons: imagine that we woke up one day (obviously suffering from collective amnesia) and had "forgotten" that the average temperature cycles throughout the year. At the first sign of sustained cold weather in November, alarms would be sounded: "The earth is growing sud-

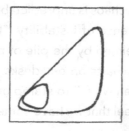

Figure 10:   Three examples of limit cycle attractors observed in the PO reaction. These plots of the system's trajectories in state space show (from left) a periodic oscillation going through a bifurcation where the period approximately doubles (center), followed by another bifurcation that quadruples the period (right).

denly colder! We need to do something to counteract this trend!" It's probably a good thing that we don't have the technological capability to do anything about plunging temperatures over the northern hemisphere every fall and winter, because to do so would wreak certain ecological disaster.

It's easy, in this example, to see how ludicrous it would be to consider interfering with the natural cyclic processes underlying seasonal temperature changes. However, do we have the same sort of insight into other cyclic processes which may be going on around us, either in our lives or in the dynamics of our organizations and businesses? The natural tendency is to try to control these fluctuations, but others prefer to let nature take its course, as is apparent in the controversy over wildlife management or control of wildfires. We *know* the cycling of seasons is a stable phenomenon, but do we believe the cycling of our inner seasons is equally stable? Do we try to control our inner seasonal fluctuations the way wildlife management in the parks[14] tries to control predator populations? By eliminating those factors we deem "bad" we may wreak havoc in our own inner ecological system.

So, it seems as if we *desire* stability — equilibrium or repetitive cycling — but, after a little reflection, I'm sure most of us would agree that this is *not* the way we would like to live forever. Even though cyclic, completely periodic behavior is more exciting than stasis, it is still pretty deadly and would be a clock-like boring existence. However, the alternative — unpredictable, irregular, even chaotic behavior — is clearly something to rid ourselves of. Isn't it? Isn't it our duty to create organizations and businesses which follow predictable patterns, in which individuals and teams can plan for the future, can know what to expect? Even if the human race has yet to create a single organization that is like this, isn't predictability at least something to *strive for* and chaos something to avoid? After all, chaos is unpredictable, irregular — chaos sounds like it would be the very essence of instability. "Utter chaos," we say as we throw our hands up, overwhelmed by the pile of papers, memos, phone messages and plummeting sales charts on our desks. Chaos is clearly the enemy and order is its opposite; when we fail to stamp out chaos in our lives, organizations or governments, we feel that we have somehow failed to achieve order, to do what is our duty.

It turns out, though, that chaos, too, is quite stable and contains a hidden order. Chaotic behavior is also governed by an attractor. It is this attractor, a strange one indeed, that assures us of the stability of even chaos and reveals the order within it. One of the great lessons of chaos theory and complexity science is that *unpredictable* and *unstable* do not mean the same thing — we *can* have one without the other; life can be unpredictable *and* stable, all at the same time.

Endnotes

1. Jonathan Star and Shahram Shiva, translators, *A Garden Beyond Paradise: The Mystical Poetry of Rumi*, Bantam Books, New York (1992), p. 102. Permission from Jonathan Star to reprint this and other short excerpts is gratefully acknowledged.

2. William Proxmire, Democrat, served as a United States Senator of Wisconsin from 1975–1988. He issued 168 "awards" monthly to both scientific projects and government agencies he thought were wasteful. The name is a play on the Order of the Golden Fleece, a chivalrous honor extending back to the late 15th century, using a double entendre for the word "fleece".

3. For information about this and other oscillatory chemical reactions, see: Richard J. Field and Maria Burger, *Oscillations and Traveling Waves in Chemical Systems*, John Wiley & Sons, New York (1985).

4. See: Albert Goldbeter, *Biochemical Oscillations and Cellular Rhythms: The Molecular Bases of Periodic and Chaotic Behavior*, Cambridge University Press, Cambridge (1996).

5. Arthur Winfree, *The Geometry of Biological Time*, Springer, New York (1980).

6. Erik Mosekilde and Ole Mouritsen, eds., *Modeling the Dynamics of Biological Systems: Nonlinear Phenomena and Pattern Formation*, Synergetics, Springer-Verlag, Berlin (1995).

7. (a) Jorge M. Davidenko, Arcady V. Pertsov, Remy Salomonsz, William Baxter and José Jalife, "Stationary and drifting spiral waves of excitation in isolated cardiac muscle," *Nature*, Vol. 355, pp. 349–351 (1992).

   (b) Leon Glass, "Dynamics of Cardiac Arrhythmias," *Physics Today*, Vol. 49, pp. 40–45 (1996).

8. A few examples from the literature:

   (a) A.H. Cornell-Bell and S.M. Finkbeiner, "Ca^{2+} waves in astrocytes," *Cell Calcium*, Vol. 12, pp. 185–204 (1991).

   (b) G. Dupont and A. Goldbeter, "Oscillations and waves of cytosolic calcium: insights from theoretical models," *BioEssays*, Vol. 14, No. 7, pp. 485–493 (1992).

   (c) M.E. Harris-White, S.A. Zanotti, S.A. Frautschy, and A.C. Charles, "Spiral intercellular calcium waves in hippocampal slice cultures," *Journal of Neurophysiology*, Vol. 79, pp. 1045–1052 (1998).

(d) L.F. Jaffe, "Classes and mechanisms of calcium waves," *Cell Calcium*, Vol. 14, pp. 736–745 (1993).

(e) L.S. Jouaville, F. Ichas, E.L. Holmuhamedov, P. Cmacho and J.D. Lechleiter, "Synchronization of calcium waves by mitochondrial substrates in Xenopus laevis oocytes," *Nature*, Vol. 377, pp. 438–441 (1995).

(f) Y.-X. Li, J. Rinzel, J. Keizer and S.S. Stojikovic, "Calcium oscillations in pituitary gonadotrophs: comparison of experiment and theory," *Proceedings of the National Academy of Science*, Vol. 91, pp. 58–62 (1994).

(g) E.A. Newman and K.R. Zahs, "Calcium waves in retinal glial cells," *Science*, Vol. 275, pp. 844–847 (1997).

(h) M. Wilkins and J. Sneyd, "Intercellular spiral waves of calcium," *Journal of Theoretical Biology*, Vol. 191, pp. 299–308 (1994).

9. The model, developed in the 1920s, is discussed in detail in his later book: Alfred J. Lotka, *Elements of Mathematical Biology*, Dover Publications, New York (1956).

10. Vito Volterra, "Variations and fluctuations of a number of individuals of animal species living together (translated from the original Italian, 1926)," in *Animal Ecology*, R.N. Chapman, ed., McGraw-Hill, New York (1931), pp. 409–448.

11. Rolf O. Peterson, "Wolf-Moose interaction in Isle Royale: The end of natural regulation?" *Ecological Applications*, Vol. 9, No. 1, pp. 10–16 (1999).

12. Willard G. Oxtoby and Roy C. Amore, *World Religions: Eastern Traditions*, 3rd Edition, Oxford University Press, New York (2010).

13. This word — equilibrium — is tricky, since it can be defined several ways, depending on whether we're considering its mathematical definition, thermodynamic definition, the way it is used in economic theory, or just ordinary speech. Here I am using the ordinary speech version of the word.

14. (a) R. Gerald Wright, "Wildlife management in the national parks: questions in search of answers," *Ecological Applications*, Vol. 9, No. 1, pp. 30–36 (1999).

(b) Michael E. Soulé, James. A. Estes, Joel Berger and Carlos Martinez del Rio, "Ecological effectiveness: conservation goals for interactive species," *Conservation Biology*, Vol. 17, No. 5, pp. 1238–1250 (2003).

Insight

5

Chaos

*In all chaos there is a cosmos, in all disorder a
secret order...*

Carl Jung[1]

Life can be chaotic at times. We have all experienced such times, but even though we know intuitively that all chaos cannot be prevented, most of us act like it can. We generally want it to stop, and wish that some semblance of order would return to our lives when things get a bit too crazy to bear.

One of the more exciting and surprising results of the scientific study of complex systems is that while chaos is very common and definitely leads to unpredictability, it can also bring with it all sorts of good things — greater creativity, an enhanced ability to adapt to change, the wisdom and comfort that comes with letting go of the need to control the uncontrollable, and other unexpected benefits. An understanding of chaotic dynamics can help us to recognize, and ultimately accept, the good things that come along with the inescapable chaos of life.

Chaos is a type of aperiodic, but still roughly cyclic, behavior that arises when a system that has already undergone a series of bifurcations, usually ones that produce cyclic states like those I discussed in the previous chapter, bifurcates once more. The result is a wholly new type of dynamic with intriguing properties — and one such property is that chaos renders the system unpredictable. This is not necessarily bad, since predictable behavior can, after all, be pretty boring.

Another property of chaos is that it's surprisingly stable. The combination of unpredictability and stability is what made the concept of chaos so intriguing to me when I first learned about it. My introduction to the concept of chaos was gradual, as I'd been reading about it for a few years and hearing about it at scientific meetings, but it wasn't until it appeared in my own lab that I really started to pay attention.

Chaos in the PO Reaction

We were carrying out computer simulations of the horseradish peroxidase-oxidase reaction at the time,[2] and many of the data sets seemed to be erratic and unpredictable. To follow the progress of the reaction, we calculated the dissolved oxygen concentration to compare to experiment. My collaborator, a Danish biochemist named Lars Folke Olsen, had carried out numerous experiments in which he followed changes in the oxygen concentration over time. Oxygen entered the solution from the atmosphere, was consumed by the enzyme-catalyzed reaction, then regenerated. The measured value of oxygen in Lars's experiments would rise, then fall, then rise again, sketching out a perfectly rhythmic and periodic wave. Our calculations, using a suggested mechanism for the reaction, showed the same thing: a periodic oscillation, such as those shown in Figure 8 in the previous chapter.

We expected to see periodic cycles in oxygen level. They were, indeed, the reason we were studying the horseradish reaction, and most of our previous experiments had behaved this way. The oscillations occurred when the reaction system experienced a bifurcation, transforming the horseradish system's point attractor into a form that we, and everybody else in the field, referred to as a limit cycle attractor (see Figure 10 in previous chapter).

In some of Lars's experiments, though, the oscillations in oxygen level were not regular or periodic at all but, rather, seemed almost totally erratic and aperiodic. In these weird experiments, the upswings in oxygen concentration would sometimes go very high, but at other times they would be only tiny blips, producing a pattern on the chart recorder that looked a bit like a stock market graph. None of us could predict where the system would go from one moment to the next. Lars would run the experiment again, the same way, and get a different result. We saw the same erratic oscillations in our computer simulations. We had a suspicion that what we were seeing was chaos, but to confirm this we needed to carry out certain mathematical tests.

The tests involved graphing the oxygen level data for the unpredictable oscillations the same way we had graphed the data for the rhythmically periodic experiments. We found,[3] as suspected, that the limit cycle attractor had changed. It had, in fact, bifurcated many times. We could immediately tell that our system, which had once been governed by a simple cyclic attractor, had undergone a cascade of bifurcations, producing something qualitatively different: a cycle, shown in Figure 11, that never precisely repeated itself. Despite the fact that the dynamics were wildly unpredictable in their precise details, they traced out an attractor that was quite similar in shape to the limit cycle attractors we'd found for periodic behavior in this reaction. This object, a rather strange one, still functioned as an attractor for the system and kept the wild out-of-control swings in oxygen level under some semblance of control.

We had found what those in the business refer to as a strange attractor, so-called because it is not a one- or even two-dimensional object, but is best described by a fractional dimension. A one-dimensional attractor would be a simple curve, like those shown in Figure 10. Our attractor was, in fact, a fractal with a dimension between 1 and 2. Because this attractor governed the chaotic behavior we were observing in this chemical reaction, we knew that our chemical system was stable, even though the chaotic nature of its trajectory meant we could not predict with much precision what it might do from one moment to the next.

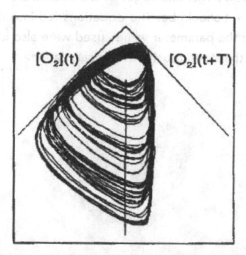

Figure 11: Experimental data for the PO reaction showing the existence of chaos and a fractal attractor. Notice the similarity in shape to the limit cycle attractors shown in Figure 10. The data is graphed using a technique known as a time-delay reconstruction. In this technique, oxygen levels at time t are plotted against oxygen levels at a later time, $t + T$.

Chaos in a Weather Model

Our attempts to make sense of the chaotic data in both Lars's experiments and our computer simulations were greatly enhanced by the work others were doing around the world. Chaos had been detected in many systems of different types and had been known since the early 1960s when it was first discovered[4] by Edward Lorenz, a professor of meteorology at MIT, in a weather model he was studying; see Figure 12.

Lorenz found[5] that his computer program for simulating weather patterns gave different results every time he ran it. In order to carry out very long simulations, Lorenz and his assistants had used a standard computer-modeling technique, one I have also used many times in my own work: they took values found on the printed output from the middle or near-end of a long simulation and used them as the starting values for the next simulation. This allowed them to chop long runs into smaller, more manageable jobs. People had used this technique for a long time and it always worked beautifully — until Lorenz and his assistants attempted it with his model.

The problem he and his assistants soon encountered was that two subsequent simulations never agreed with each other. Every time they used values from the middle of his previous output to start a new run, the numerical results of the second run did not agree with those of the first. Small discrepancies between the two grew larger and larger as the second simulation progressed.

This was not how it should be. The equations fed into the computer each time were identical; the parameter values used were also identical. All Lorenz had done was shift the starting point of the calculation.

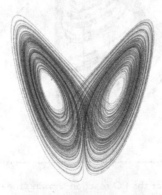

Figure 12: The Lorenz attractor for parameter values associated with chaos in the model.

Everything he knew of how numerical modeling was supposed to work said that each computer run should give identical results, and yet, the behaviors of the two simulations were very different. Of course, he thought there was a bug (who wouldn't?) but after many tests of the software, soon realized that the program was bug-free and that what his group had stumbled upon was actually more interesting than a mere mistake. Even though they had specified the system *exactly*, no one could predict, even approximately, how it would behave from one run to the next. This unpredictability went against everything Lorenz knew about how Newtonian physics was supposed to work.

Lorenz came to the conclusion that what they had found was a genuinely new phenomenon in science, a result that was totally unexpected and, indeed, believed to be impossible before they did their calculations. The idea that a completely determined system could give seemingly random predictions of the future seemed to go against what scientists had always thought — always being about 400 years. Isaac Newton himself had said that once the equations of motion for a system were determined and the parameters specified, solutions of those equations should be the same no matter *how* many times one solved them. Newton said it was these equations that determined the future course of events and once one knew what the equations were, this knowledge should, in principle, be all that we would need to calculate the future behavior of that system for all time.

This clockwork universe, as Newton's ideas had become known, were so firmly entrenched in scientific thought by the time Lorenz and his assistants ran headlong into chaos that they were no longer mere hypotheses or theories — they were *law*. And Newton's law implied this: if a system can be described exactly (i.e., it is *deterministic*), its future is fixed (i.e., *determined*) and can be calculated. Lorenz's result showed that this was not true: even a deterministic system, he showed, can have an unpredictable future, no matter how precisely the system is understood.

The result from Lorenz's weather model changed everything. Although Newton's laws were technically only applicable to material objects acting under the influence of physical forces, even religious thinkers took them to heart. The idea that the universe's actions should be completely predictable if one had total knowledge of every condition led to a host of religious beliefs that fit this scenario. The concept of pre-destination is a prime example of this reasoning. If God is omniscient, thus knowing everything about every part of the universe, how could He not also know precisely what will happen to each of us in the future? This view is based entirely on Newton's ideas of a clockwork universe. The discovery of chaotic behavior that comes about even in a perfectly well-defined system raises the possibility that even an omniscient God might not be able to know the future.

It is now accepted by many investigators in the field that Lorenz was not the first to observe chaotic behavior. This distinction goes to Henri Poincaré, the French mathematician who laid the foundation for the geometric understanding of planetary dynamics. As recounted by Ian Stewart in his book[6] *Does God Play Dice?*, Poincaré was unable to prove that the solar system was stable. He was trying to prove this in response to a contest established by the King of Sweden, but failed. In hindsight, many of Poincaré's solutions of the equations describing planetary motion look remarkably like what we would now call chaos.

Furthermore, recent evidence has surfaced[7] that Lorenz was assisted in his work by two women, Ellen Fetter and Margaret Hamilton, both computational experts. At the time, women carried out many of the tedious computations in science, but rarely received credit or co-authorship. Lorenz thanked both women in the acknowledgements for his publications, but he did not extend co-authorship to either one.

Some people have claimed that a second breakdown in theology occurred with the advent of quantum theory, specifically the Heisenberg Uncertainty principle.[8] In simple terms, the principle says that it is not possible to exactly and simultaneously know both the momentum and position of a particle. If one of these is known precisely, the other can only be estimated. There is, then, a degree of uncertainty in our ability to precisely describe the state of the system we are studying at any given moment.

The Heisenberg principle has meaningful consequences only for small objects subject to the laws of quantum mechanics. The behavior of a large group of particles brought together into a massive object such as a human being or even a baseball will not be affected by this uncertainty. This has not stopped people from being philosophically drawn to the idea, however. The very idea of an uncertainty principle seems to bring into question how much we can actually know about the universe we live in. The discovery of chaotic behavior in a deterministic system brought a separate type of uncertainty into our understanding of the universe.

Chaotic behavior arises in large systems, such as our brains, the atmosphere, bodies of water, and possibly even societies — systems typically comprised of billions, trillions, or even more atoms. Unlike quantum mechanics and the Heisenberg principle that apply only on the atomic or subatomic level, the type of chaos exhibited by Lorenz's model can arise in the kinds of systems we live in and deal with on a daily basis, such as the weather, the behavior of a flock of birds or a crowd of people, even the functioning of biochemical reactions in our body.[9] Since Lorenz's discovery and the many subsequent studies that confirmed the existence of chaos in other settings, many people have become familiar with the basic ideas of what has come to be popularly known as chaos theory.

One of the properties of systems that display chaotic behavior (like Lorenz's weather model) is a phenomenon known as sensitive dependence on initial conditions. This was famously referred to in the movie *Jurassic Park* as the butterfly effect. It describes how a small perturbation, such as a butterfly flapping its wings, can be amplified into a major event. Through this effect, the weather in one part of the world (the US, say) can be changed by as small a perturbation as the tiny shift in wind caused by a fluttering butterfly in another part of the world (say Australia). The small fluctuation in wind speed and velocity caused by the butterfly's wings is amplified by the nonlinear processes occurring in the atmosphere into a large effect — maybe even a hurricane, although the result might just as well be a beautiful high-pressure system with sunny skies.

It was, in fact, the butterfly effect that caused the chaos in Lorenz's original computer experiments. In his case, though, there was no real butterfly but there was a small perturbation in the starting conditions for his simulation. Lorenz's butterfly was, in fact, the shortcut his group had devised. When his assistant took a data point from the printout in the middle of a simulation run and used it as the initial point to start another simulation, the two simulations did not agree after several iterations. This was because the starting value on the printout had been rounded off from the numerical value found in the internal computer register. The printed value was actually slightly different from the stored value in the computer's registers, so their shortcut had introduced a perturbation similar to that made by a small flap of a butterfly's wings.

Lorenz tried increasing the precision of the printed numbers, but this didn't help; no matter how small the difference between the number stored in the computer and the number printed out, that difference quickly became amplified as the new simulation proceeded. The result was a model that, although completely known and perfectly defined, could never precisely predict the weather. It turns out that all weather models are like this, not just the one

Lorenz studied that year, and for this reason we now know that it will *never* be possible to predict the weather for a long period of time.

Many more examples of chaotic behavior have been observed in laboratories around the world, including our own, since that day Lorenz first stumbled upon the phenomenon of chaos. My colleagues and I found chaotic behavior in the PO reaction as I described earlier, but it has also been observed in fluid flow, other chemical reactions, mechanical systems, biological systems — even in financial and economic systems. The idea that systems like the one Lorenz was studying (completely specified equations describing a few simple parts interacting by equally simple rules) could display unpredictable behavior, or chaos, has since risen from a hypothesis to a theory — maybe not quite yet a *law*, but the existence of chaos in many different types of systems is considered fact now.

Many feel that the discovery of chaos was a true paradigm shift, a modern-day illustration of Thomas Kuhn's theory of scientific progress. James Gleick called it "The New Science" in his book[10] that described the studies of some of the people who worked out the initial ideas about chaos and how it functioned. When chaotic behavior was first characterized, the prevailing paradigm shifted. The existence of chaos in a completely determined system showed that Newton was wrong: the universe is not a giant clock which, once wound up, will head toward a predetermined future. Since the discovery of chaos, we now know that the future for even very well-defined systems can never be known precisely — no matter *how* well we understand them.

Drawing Life Lessons from Nature

Long before I became a card-carrying scientist, I drew lessons about my life from nature, so it didn't seem odd to me to consider whether my own life dynamics were similar in some way to the dynamics of the chemical reaction in horseradish I was studying. A lot of people draw sweeping lessons about human behavior from nature, even when they don't admit it or, perhaps, even realize it. For example, the Darwinian concept of survival of the fittest is behind a lot of the arguments for why the free enterprise system is the most natural economic system. Nobel Prize-winning economist Elinor Ostrom has shown,[11] though, that this is not always the case. Her work revealed the importance of cooperation, as opposed to competition, in determining people's economic choices. Her conclusions would not surprise ecologists, however, who know that cooperation is often the key to stabilizing an ecological system.

We often assume that people behave the way nature does. This isn't so far-fetched, since we are a part of nature, after all. The problem with basing one's conclusions about what people are naturally like on what science tells us nature is like is that we don't fully understand nature yet. Those who assumed that "Nature red in tooth and claw"[12] was the fundamental law governing inter-actions of species were forced to change their view by new insights into sym-biosis and adaptive behavior in ecological systems. Darwin's original concept of survival of the fittest is just the first part of what we now understand as the natural way species exist in an ecosystem.

Science keeps uncovering new and surprising things about the world that were previously thought to be impossible, like cooperation between species that ought to be competing. Take the surprising relationship between the clownfish and the sea anemone among which it lives. The anemone's stinging tentacles would normally be toxic to the fish, but the clownfish has evolved special mucus to protect it from the stings. The clownfish, in turn, protects the anemone from other predatory fish, while the stinging tentacles of the anemo-ne protect the clownfish from its own predators. This degree of cooperation is not unusual. Many examples of symbiotic or near-symbiotic behavior are now known from lichens to pilot fish and sharks.

Chaos is another thing that science since the time of Newton had taught us was supposed to be impossible: a perfectly well-defined system should behave in a perfectly predictable fashion, just as competing species should not coop-erate. The trouble with this 400-year-old view is that it is just not true — not for the universe and not for our lives.

So, what are we, as individual persons, to make of this insight from the new science? Does it mean the universe and everything in it, including our own lives, is unpredictable and essentially chaotic? Life *does* seem to be basically unpredictable — but is chaos the right term for that? Science seems to be con-firming what some of us feared: that there is no way to know (or control) what will happen in our lives. Worse, these insights seem to imply that *nothing* is in control and whatever happens just, well, *happens*.

Chaos theory raises many questions when we try to apply it to our lives. Are the twists and turns of our individual lives merely random selections out of the many possibilities that *could* have happened? Is there no *meaning* to the tragedies we endure? Or to the joys? If we live a long, healthy, happy life or a short, painful, tragic one, is this just a random event, the result of a dice roll? Is it the case that the tragic (and happy) events from which we so often try to draw lessons about our lives are inherently meaningless?

Chaos theory, as it is usually portrayed, seems to say that none of the events of our lives — the death of a loved one, the joy of an unexpected success, a senseless murder, an apparently meaningful chance encounter — were ever *meant* to happen. However, if all we focus on is the chaos and pay no attention to the underlying attractor, this depressing conclusion might be where we end up. Knowing about the attractor can bring us to a completely different conclusion, one that promises not only comfort and security, but also excitement, surprise and joy.

Order in Chaos: The Strange Attractor

Before I learned about chaos through my work in science, I thought, like everybody else, that a chaotic life was something to avoid. By the time I was embroiled in my own bifurcation event, when my life had become about as chaotic as it ever had been up to that point, I had a different perspective on chaos. I knew all about the butterfly effect, the sensitive dependence on initial conditions, the unpredictability of a chaotic system — but these were not what intrigued me about chaotic behavior. What I kept thinking about was the *attractor* — the beautiful underlying order beneath the chaos. The attractor was evidence of the stability of the system even in the face of full-blown disorder. Without knowing about the attractor, chaos may very well seem like something we ought to stamp out. With a knowledge of the attractor, though, we might surprise ourselves by welcoming chaos into our lives.

Most popularized versions of chaos theory miss the most interesting part of the theory, which is the existence of the attractor, something that Lorenz saw right away, but many of the popularizations of the theory have ignored or not fully appreciated. As we have seen, this mysterious and hard to understand mathematical object undergirds and governs many nonlinear systems — including those that behave chaotically. The attractor is *the most important part of chaos theory*, but very few casual observers have a good understanding of it — if they have even heard of it.

If your life is as chaotic as mine has been, perhaps a knowledge and understanding of the attractor will bring with it some comfort — because even when we can't see or sense it, the attractor is always there, stabilizing any chaotic system. This assurance might bring some peace. Chaos is unpredictable, that's a given; what it isn't, though, is *unstable*. And that's the whole point of the theory known as chaos theory: while Newton can no longer tell us exactly what the future holds, the theories of chaos tell us that the unpredictable chaotic behavior characteristic of so many nonlinear systems is always stable.

Many nonlinear dynamic systems evolve through a sequence of bifurcations to more and more complex states and the ultimate culmination of this bifurcation sequence can be a chaotic state. This is, in fact, what happens in the peroxidase-oxidase reaction: the limit cycle attractor goes through a cascade of bifurcations that each double the period until it, ultimately, becomes infinite.

When chaos is finally arrived at, the attractor has bifurcated so many times that it has taken on a special geometric form known as a fractal. A chaotic attractor is also known as a strange attractor, because of this unusual geometry. Fractals, which we will discuss more fully in the next chapter, are self-similar, meaning that each part is a reflection of the whole. So, interestingly, when chaos exists, the attractor that stabilizes the chaos takes on a self-similar form — one in which each part is a copy of the whole, promising insights so profound they might even entice a person to seek out chaos. In the next chapter we will explore, more deeply, the spiritual implications of a self-similar attractor, but suffice it, for now, to say that Carl Jung had it right when he said, "In all chaos there is a cosmos."[1]

Walking the Labyrinth

The statement, "Life is chaos," should really be, "Life is chaos, but a knowledge of the attractor will allow us to accept the unknowability of life. The attractor assures us that life is stable — no matter how unpredictable the chaos may become." There is an order within chaos — and this order is, quite literally, the attractor itself.

Even though I had a good grasp of this concept before I encountered my own personal bifurcation, I didn't completely appreciate what it meant for me until I was plunged into my own chaos. Lost in confusion for months, I finally saw that an attractor did, indeed, exist and that it had something to do with my love of writing. This allowed me to trust that the chaos I was living was actually a sign of growth, not the disaster I'd feared.

Knowledge that there was an underlying attractor gave shape and form to the chaotic nature of my life. The attractor was impossible to see or sense as long as I was focused on the details of my life, unable to see past the current week or day — often unable to see past the current minute. Sensing the attractor required an ability to get outside of myself, to see my life whole the way the Apollo astronauts saw our world whole for the first time on their flight to the moon.

It is often necessary to get outside of a system to glimpse its wholeness; similarly, it is sometimes necessary to stand back from ourselves to see that *we are whole* and that our lives are working just fine. I had been looking, for months, in my journal entries for evidence of an attractor, but it was not until I participated in a ritual that allowed me to *physically sense* the attractor that I truly understood it.

The ritual was of the circumambulation type, similar to the Hajj in one sense since, like the Hajj, its origins were in an ancient ritual pilgrimage. Popular among medieval Christians, it involved walking a labyrinth, an elaborate design generally carved into a cathedral floor. This type of walking meditation has experienced a revival of popularity in recent years, although it generally has nothing to do with full-blown pilgrimages nowadays.

Through the leadership of Grace Cathedral in San Francisco, copies of the 13th-century labyrinth carved into the stone floor of the cathedral in Chartres, France, are being created around the world in stone, painted on canvas or plastic, even mown into grass or planted as hedges. The labyrinth design, shown in Figure 13, is referred to as a unicursal, or single-path, design since it contains no dead ends or decision points as a maze does. There is one path through the labyrinth and it inevitably leads into the center; the way out from the center is found by retracing, in reverse order, the same path.

The point of walking a labyrinth, then, is not to solve a puzzle, as one might do when finding their way through a maze. Rather, walking the labyrinth allows one to let go of the need to solve puzzles for a while, to give up being in control or having to figure everything out. The original labyrinth in Chartres provided a ritual ending to the pilgrimages that good medieval Christians were required to make to one of the cathedrals spread across Europe. Just as Muslims still do in the Hajj, these Christians were expected to make a pilgrimage, at some time in their life, to Jerusalem. With the outbreak of the Crusades the pilgrimages

Figure 13: The labyrinth design used at Chartres cathedral.

became very dangerous and the church in Rome established several alternative Jerusalems at cathedrals across Europe; Chartres was one of these.

Although we don't know for certain what these medieval pilgrims felt when the goal of their journey was reached, it was certain to have been a peak experience after so many miles of traveling. Even without the preceding long trip, though, one can recapture the sense of reaching the goal of one's pilgrimage by simply walking the labyrinth itself.

Lauren Artress, the director of the Labyrinth Project at Grace Cathedral, describes the experience of traversing the labyrinth in her book *Walking a Sacred Path* this way:

"The metaphors within the labyrinth are endless because they are shaped by our creative imaginations. Most immediate are the journey to our center of being and the creation of order from chaos. Completion, competition, emptying, turning our back on the center, distrusting our judgment — whatever our psyches need to deal with becomes the spiritual lesson of the labyrinth. ... As soon as you get settled into the labyrinth walk and get your bearings, one or more metaphors may spark within. The walk, and all that happens on it, can be grasped through the intuitive, pattern-discerning faculty of the person walking it. The genius of this tool is that it reflects back to the seeker whatever he or she needs to discover from a new level of awareness. When the ego is not tightly engaged in control, it joins the other parts of our being to allow us to see through the moment, to see beyond ourselves into the dynamic that is unfolding before us."[13]

I first had the opportunity to walk the labyrinth at a local church that had created a copy of the Chartres labyrinth using a starter kit from Grace Cathedral. At that point, I'd heard of labyrinths, but didn't know a thing about them.

After finding my way to the lobby, I settled in with a few other visitors, none of whom I had ever seen before, to listen to an explanation of the ritual. The woman who welcomed us that day explained that the walk experience typically has three stages. The first stage occurs as one winds in toward the center of the labyrinth and is a time of purgation, of letting go of the concerns of the world and of focusing inward. When the center is reached, the second stage begins: this stage consists of *not* moving through the labyrinth, but, rather, becoming still for a while to reflect on the first stage of the walk — or, to just let happen what might. One may stay in the center, she said, as long as one likes and experience whatever occurs there, if anything. In the third and last stage, the walker winds out from the center, following the same path taken on the way in, but in reverse. The third stage is, thus, a chance to reflect on how one might take what has been encountered on the walk or found in the center out into the world.

I followed the others into the room where the labyrinth was laid out on the floor. I walked for a long time, perhaps an hour, following the path that was clearly marked and easy to see, but I was never able to see more than 10 or 15 feet in front of me and could not predict when the path would make a sudden turn. One moment I was near the center, and I felt excited, thinking: *I'm almost there! There it is!* But a moment later the path had taken me back to the outside circumference of the circular design, as far from the center as one can get. I never knew if I was near the end of the first stage or not, and I soon stopped trying to predict how much longer I would have to walk.

It was impossible to tell how far I had come along the path and, yet, it was equally unlikely that I would get lost as long as I kept my eye on the path in front of me. I was aware, through my peripheral vision, of other people in the room, moving slowly this way or that, occasionally passing quite near, but I never looked up to meet another person's gaze and I was only tangentially aware that they were there. As I thought about these people, I was flooded with a sense of companionship. We were all in this together, walking our journeys in our own individual ways, at our own paces, and all equally unable to see very far along the path stretching out in front of us. The best any of us could do was just keep moving ahead and trust that the path would get us where we needed to go.

I felt a strong surge of gratitude for these companions, people I didn't even know. It was good to have fellow travelers on my journey, something I had definitely *not* felt much before that moment. My mind began to wander over the puzzling events of the previous few years, the pain I'd been experiencing as I contemplated leaving my life in science behind, the uncertainty of the future. My children were growing up and would soon leave home for college and beyond. What did this mean for me? So much of my life had centered on raising children — for years. I was entering foreign territory in many aspects of my life.

I soon forgot where I was as I became lost in my thoughts but continued to put one foot methodically in front of the other. Walking the labyrinth had become an automatic activity that I was now able to do without thinking much about it. And then, before I realized what was happening, I was walking directly into the center. I had arrived, but I hadn't even known I was close — it happened as soon as I stopped thinking about getting there.

I felt tired from the long, slow walk, but also a little sad or disappointed. Where was the revelation that was supposed to hit me now that I had reached the goal? Where was the insight I was supposed to find after walking all this time? Feeling impatient, I stepped back on the path to walk out of the center.

As I traveled the labyrinth in the reverse direction, I continued to feel impatient and agitated. Maybe if I had thought a little harder about it, I would have recognized the impatient feeling for what it was — the unease that descends just prior to a creative act or a big breakthrough. This has happened to me often enough that I ought to recognize the feeling — but I never do, until the insight has revealed itself. I found it increasingly difficult to think as I walked, so I eventually stopped thinking, and just walked.

Moving along the path was easy enough: put one foot in front of the other; repeat. The twists and turns of the path were still a mystery and came upon me by surprise, but this time it didn't concern me. I no longer needed to know all those details; this time I had the assurance that the path would lead me to where I needed to go, even though I couldn't see more than a few feet in front of me to prepare for sudden turns. I liked this feeling of not having to be in charge, not having to be alert and on guard all the time. I knew that the pattern of the entire labyrinth was much more than I could hold in view, much bigger and more elaborate than I could hold in my mind at one time, but I also knew that I could trust it to get me where I needed to go.

I reached the end of the path and stepped out, onto the wooden floor; as soon as my foot left the plastic sheet and touched the wood planks, the truth about the labyrinth hit me. I whirled and looked back at the path I had just finished traversing; it wound around and around in a complex pattern, one I still could not figure out just by looking at it. *The attractor is like that, too,* I realized. It's a convoluted, complicated pathway that cycles around and around a center with twists and turns that cannot be predicted — but the overall pattern is orderly and stable. The stability prevents anyone traveling the path from ever flying completely away from the center — even though we can't know precisely what lies ahead for us. The attractor holds it all together, even if I, as an individual, cannot see the whole pattern, cannot figure out how it works. The labyrinth, in other words, is a physical re-creation of the attractor — one upon which I could actually walk, thus participating in a ritual in which I experience the twists and turns of my own attractor, but eventually come to know its stability, the force that holds it all together.

Perhaps the labyrinth walk has become so popular for precisely this reason — walking it counteracts our modern tendency to try to figure everything out and, in doing so, to be in control of our lives. It's a false sense of control, of course, and the labyrinth allows us to physically experience letting go of our need to control our destiny. It reveals that something much bigger and more elaborate than we can ever fathom with our rational mind *is* in control of our lives, even if we, personally, are not.

The labyrinth provides an image of underlying stability — an image of something akin to the attractor. Physical circumambulation rituals, particularly those carried out in a meditative way, such as a labyrinth walk, the Hajj, or circling around a stupa can bring us into contact with our own attractor. And if, as was the case for me at the time I walked my first labyrinth, our attractor has bifurcated to the point where we are now in chaos, it will be a very strange and wonderful attractor we will encounter by participating in a walking meditation at such a special time. The insights that can be revealed through this simple movement meditation tool are often profound at times of great turmoil. And in the process of coming to know this strange, beautiful attractor, we will also come to know that, far from being small, helpless beings, buffeted by the forces of a chaotic life, we are actually, deep down, powerful beyond measure.

Stability in the Presence of Instability

Another thing that is different about chaotic dynamics, as opposed to the simpler periodic oscillations or cycles that we considered before, is that chaos is governed by two opposing effects or forces. One force tends to push trajectories apart from each other while the other force pulls them back together again, keeping everything moving in the same general direction even in the face of relatively uncontrolled changes at arbitrary times due to the repelling force. These two opposing forces — one repelling, the other attracting — reflect the influences of two *manifolds*, or sheets, in the underlying landscape, each with opposite influences on the trajectories. These two sheets create a stable direction and an unstable one.

This is because the sheets form a surface in phase space that is like a mountain pass: a saddle point. If you imagine standing on the surface at this point, it would slope downward to the right and left, but slope upward in front and behind you. A slight step to the left or right would make you fall downward — this is the unstable direction. However, stepping forward or back would require a climb — this is the stable direction.[14]

Trajectories can move in the unstable direction at unpredictable times giving the chaotic trajectory its characteristic nature: long stretches of nearly cyclic, almost predictable motion, interspersed with occasional wild excursions to completely different, but still somewhat cyclic, intervals.

The two opposing forces which cause this, while seemingly contradictory and opposite, exist simultaneously; both are necessary for the existence of chaotic motion in physical systems. Both are also necessary for its stability. The tension between the destabilizing and restabilizing forces is more than just a seeming paradox: it is the origin of chaotic, unpredictable behavior that is, nevertheless, quite stable.

"Divinity," Nicholas of Cusa said,[15] "is the coincidence of opposites." Carl Jung also advocated the importance of the concept of a union of opposites to psychological and spiritual health. And Scott Peck, in *Further Along the Road Less Traveled,* writes[16] that the existence of paradox, the co-existence of opposites, is a sure sign that we are in the presence of the divine. All three of these thinkers seem to have hit on the same truth, one that is hidden in the very structure of the attractor that underlies and stabilizes chaos. The co-existence of a stabilizing force and a destabilizing one is central to, indeed *defines,* chaotic motion on a strange attractor; perhaps this is also true for the chaotic attractor in our spiritual state space.

The confluence of opposing forces of the type we see in the strange attractor seems to have been intuited by Nicholas of Cusa who could not, of course, have known anything about attractors. Consider the otherwise seemingly contradictory passage: "Divinity is the enfolding and unfolding of everything that is. Divinity is in all things in such a way that all things are in divinity…I return again to the divine enfolding and unfolding. Returning, I go in and out of divinity. When I find God as the power that unfolds I go out. When I find God as the power that enfolds I go in. When I find God as the power that enfolds and unfolds I go in and out simultaneously."[17]

In his original writings, Nicholas used the terms *complicatio* (enfolding) and *explicatio* (unfolding). The former, also possibly interpreted as *becoming complicated*, was used to describe the part of the dynamic that causes us to "go in", as he says. We can also think of this as becoming more complex, more evolved. *Complicatio* is active in the process of creation. Whenever we turn inward, the step that always precedes expressing ourselves through writing, dance, art, etc., we are emphasizing this part of the dynamic of chaos.

Explicatio, which has been interpreted as unfolding, could just as well have been interpreted as *becoming explicated* or explained. This is the part of the dynamic of chaos that leads us out into the world. Or, as Nicholas wrote[18] centuries ago,

> *I go in passing from creation to Creator.*
> *I go out passing from Creator to creation.*

Endnotes

1. Carl Jung (1875–1961), *Collected Works, Vol. 9, "The Archetypes and the Collective Unconscious,"* pt. 1 (R.F.C. Hull, translator), 2nd Edition, Princeton University Press, New Jersey (1981); Permission from Princeton University Press to use this quote is gratefully acknowledged.
2. Some of our early results were published here: T. Geest, C.G. Steinmetz, L.F. Olsen, R. Larter and W. Schaffer, "Nonlinear Analyses of the Peroxidase-Oxidase Reaction," *Journal of Physical Chemistry*, Vol. 97, pp. 8431–8441 (1993).
3. T. Geest, C.G. Steinmetz and R. Larter, "Universality in the Peroxidase-Oxidase Reaction: Period Doublings, Chaos, Period Three and Unstable Limit Cycles," *Journal of Physical Chemistry*, Vol. 97, pp. 5649–5653 (1993).

4. Edward N. Lorenz, "Deterministic Nonperiodic Flow," *Journal of Atmospheric Science*, Vol. 20, p. 130 (1963).

5. Lorenz recounts his experience as he discovered the phenomenon of chaos in his memoir: Edward N. Lorenz, *The Essence of Chaos*, University of Washington Press, Washington (1993).

6. Ian Stewart, *Does God Play Dice? The Mathematics of Chaos*, Blackwell Publishing, Massachusetts (1989).

7. The story is recounted in "The Hidden Heroines of Chaos," by Joshua Sokol, Quanta Magazine, May 20, 2019, www.quantamagazine.org.

8. Walter J. Moore, *Physical Chemistry*, 5th Edition, Prentice-Hall, New Jersey (1972).

9. A good set of reprints on the topic of chaos has been collected here: Predrag Cvitanović, *Universality in Chaos*, 2nd Edition, Adam Hilger Publishers, Bristol and New York (1989).

10. James Gleick, *Chaos: Making a New Science*, Viking, New York (1987).

11. https://www.nobelprize.org/prizes/economic-sciences/2009/ostrom/facts/.

12. From the poem, *In Memoriam A.H.H.* by Alfred Lord Tennyson, now widely associated with Charles Darwin's theories.

13. Lauren Artress, *Walking a Sacred Path: Rediscovering the Labyrinth as a Spiritual Tool*, Riverhead Books, New York (1995), pp. 96–97; permission from Penguin Random House LLC to reproduce this excerpt is gratefully acknowledged.

14. For more detailed mathematical descriptions, see the following references:
(a) John Guckenheimer and Philip Homes, *Nonlinear Oscillations, Dynamic Systems, and Bifurcations of Vector Fields*, Springer-Verlag, New York (1983).
(b) Heinz Georg Schuster, *Deterministic Chaos: An Introduction*, 2nd Revised Edition, VCH Publishers, Weinheim (1988).

15. James Francis Yockey, *Meditations with Nicholas of Cusa*, Bear and Company, New Mexico (1987); permission from Inner Traditions/Bear and Company to use this and other excerpts is gratefully acknowledged.

16. M. Scott Peck, *Further Along the Road Less Traveled*, 2nd Edition, Touchstone (1998); See also M. Scott Peck, *A World Waiting to be Born*, Bantam Books, New York (1993).

17. Yockey, pp. 29–30.

18. Yockey, p. 31.

Insight

6

Fractals

The soul comes once,
this body a thousand times.
Look closely —
a million waves
One sea.

Rumi[1]

The attractor of a system that is behaving chaotically is a strange object known as a fractal. As we will see in the following sections, fractals have a geometry that is best described with a fractional dimension. The scientists who discovered chaos did not, at first, know they were dealing with a phenomenon that appears in many realms of nature, so they referred to the unusual attractor as "strange".

The strangeness of the fractal chaotic attractor is due to the fact that it has essentially the same pattern of trajectories at different levels. This phenomenon is known as self-similarity, meaning the attractor's appearance at high levels of magnification looks a lot like the entire attractor. Otto Rössler, a well-known complex systems scientist who I mentioned in an earlier chapter, discovered one simple system of equations that illustrate this property. Chaotic behavior in Rössler's model[2] is associated with a strange attractor that has the appearance of a loop circling a central point, but not quite closing the circle (see Figure 1 in Chapter 1 for an illustration of this attractor). After making a complete circuit around the center, the loop misses its own starting point by a small distance and traces out a slightly different loop on its next circuit around the attractor's perimeter. This happens each time the trajectory goes around so that, over time, the trajectory's lines begin to fill in a region with many nearly identical loops that almost, but not quite, match up.

The result is an object that is partway between a one-dimensional curve (a single closed loop) and a two-dimensional surface (a completely filled-in area). Benoit Mandelbrot, a mathematician and computer scientist, dubbed this type of mathematical object that is not-quite-a-curve and not-quite-a-surface a "fractal". He invented this name to indicate that the object is best thought of as having a dimension that is a fraction. As Mandelbrot says in his book[3] *The Fractal Geometry of Nature*, he took a great deal of care in choosing a name for these mathematical objects. "There is a saying in Latin," he writes, "that *to name is to know*."[4] Mandelbrot derived the term "fractal" from the Latin adjective *fractus*, which means broken into pieces, or fragmented. *Fractus* is the root of our English word "fracture", but also of the word "fraction".

The fractal concept comes up in nonlinear science in a deep and fundamental way since a strange attractor associated with chaotic behavior is a fractal object. Only chaotic attractors have this property — steady state and limit cycle attractors don't.

But there is a reverse connection as well. Take the Mandelbrot set, for example. This simple mathematical relationship, discovered by Mandelbrot, is widely reproduced since it produces beautiful color images when graphed in a certain way, such as the example shown in Figure 14. What is even more remarkable, though, and less widely appreciated is that the Mandelbrot relationship

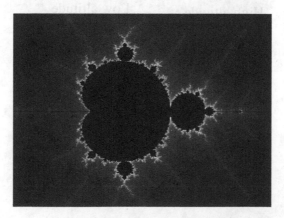

Figure 14:   A portion of the Mandelbrot set, showing its self-similar structure.

generates chaotic trajectories. The main take-home lesson here is that chaos and fractals are two sides of the same coin. The type of dynamic relationships that lead to the stable but unpredictable behavior of chaos also, remarkably, lead to the self-similarity of fractals.

The Fractal Coastline

When Mandelbrot coined the fractal term, he introduced it by posing the question: What is the length of the coastline of Britain?

At first glance, this seems like a perfectly well-posed question but, after a little thought, we see that the length of something like a coastline is a very slippery concept. We can easily imagine a simple process for measuring the length of Britain's coastline that illustrates the problem. Suppose a satellite with a camera for photographing features on the earth is put into orbit at an altitude of 20 miles and a picture of Britain is taken (see Figure 15, top left). Then, imagine that the satellite is brought closer to the surface of the earth and another photo snapped (Figure 15, top right). At half the altitude, 10 miles, more features of the coastline would be visible than at the higher altitude of 20 miles; what had appeared in the 20-mile-high photo as smooth parts of the coastline would, in the 10-mile picture, be seen to be filled with many bays, inlets or small peninsulas.

Comparing the total length of the coastline in these two pictures would reveal a longer coastline in the 10-mile altitude photo than that apparent in the 20-mile photo, simply because more features can be seen in the picture

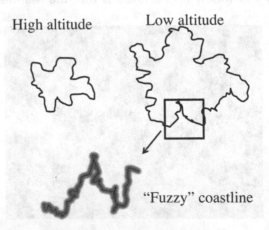

High altitude   Low altitude

"Fuzzy" coastline

Figure 15:   Illustration of the fractal coastline thought process.

taken at closer range. If this process were to be continued by bringing the satellite closer and closer to the surface of the earth, the apparent length of the coastline would continue to grow larger and larger as more twists and turns are revealed.

So, what *is* the length of the coastline? This thought experiment seems to be suggesting that the coastline length depends on the altitude from which the picture is taken. Does this mean that the true, total length of the coastline is *infinite*? This seems to be a possible conclusion of our thought experiment, since we can conceive of increasing the resolution of our photographs by moving closer and closer to the surface, all the while increasing the apparent length. However, this is, essentially a nonsense result since it ought to be possible to show that one island (say Britain) has a longer coastline than a smaller island (say the island of Manhattan). In order to compare the length of the coastline of Britain with that of Manhattan, must we compare two seemingly infinite measurements or always specify the altitude and resolution of our photos? The concept of *fractal dimension* and its associated *measure* (a fancy word for "length") will get us out of this unhelpful result.

It is useful to start by thinking of the coastline as being "fuzzy", as shown in Figure 15, bottom. This "fuzzy" coastline is composed of a two-dimensional band containing a one-dimensional curve with a large number of twists and turns in it. The twists and turns are all the bays and inlets revealed in our close-up photographs. If, rather than being fuzzy, the band were *completely* filled with these twists and turns, the one-dimensional curve would be more accurately described as a two-dimensional area. A "fuzzy" band, on the other hand, would not be completely filled by the twisting and turning one-dimensional curve, but only *partially* filled. If we consider such a fuzzy object as having a *fractional dimension*, that is, a dimension between 1 and 2, then we open ourselves to the possibility that the coastline is a fractal object whose size can be measured.

To determine the "length" of a coastline that is best thought of as having a fractional dimension, we use two concepts: *metric* and *measure*. The latter is the mathematical word for quantities such as length, area, or volume that are associated with objects of dimension 1, 2, or 3, respectively. *Measure* is, then, a general term for length, area, volume or whatever the analogous quantity is for an object with a fractional dimension. Let's call it M. The *metric* is simply the "measuring stick" used to find the actual *measure* of that fractal object. If we imagine a small measuring stick of length a and place N sticks of that size along the coastline, the total length L of the coastline is simply $N \times a$. If we try to find the area of the full island we would need to cover it with small squares, length a on each side. The full area A is then $N \times a^2$. For a fractal object, then, the measure M is given by the formula $N \times a^D$, where D is the fractional dimension.[5]

The Natural World is Fractal

I had known about Mandelbrot's work on fractals, and Rössler's as well, for a long time before I encountered the concept in my own work on the PO reaction. As Mandelbrot showed so persuasively in his book, published in 1977 when I was a first-year graduate student, fractals appear everywhere in nature[6] — in fact, it is said that the geometry of the natural world is fractal, not the Euclidean geometry of triangles and cubes we learn about in school.

I remember how much more beautiful the sight of bare tree branches against a gray November sky became to me once I had learned to see them as natural fractals. Trees branch out in a self-similar geometry, as shown in Figure 16. Each large branch can be seen to trace out the same basic pattern as the entire

Figure 16:   Examples of self-similar branching patterns in a variety of trees.

tree; each smaller branch has the same shape as the larger branch and, in turn, also the same shape as the entire tree.

If it's not yet winter as you read this book, there's no need to wait for the leaves to fall from the trees to find fractals in nature. Ferns provide another excellent example of a natural fractal (see Figure 17). All along the long fern frond run rows of leaves, each leaf a perfect copy of the entire fern. And each of these leaves is composed, in turn, of smaller leaflets, each also a copy of the entire frond.

Perhaps my favorite example of a natural fractal is Romanescu broccoli, shown in Figure 18, which looks like a cross between broccoli and cauliflower. In this remarkable vegetable, available at many farmer's markets or even your local grocery store, the individual florets are spirals, each arranged in self-similar spiraling rows that make up the head of broccoli.

The same type of self-similarity can be seen in a host of natural phenomena: bronchial tubes in our lungs, frost growing on a window pane, lightning strikes, and many other places. While the fractal concept applies in a literal way to the geometry of the natural world, it has also been used in recent years to develop algorithms for making computer-constructed images of natural landscapes that are very realistic in their appearance.

Figure 17:   Two views of fern frond, showing its self-similar structure. The left image is a computer-generated drawing using the mathematical properties of a natural fern, seen on the right.

Figure 18:   Romanescu broccoli floret showing its self-similar structure.

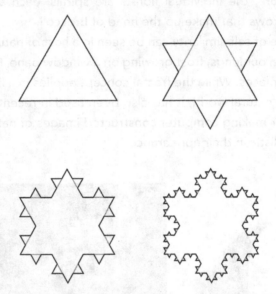

Figure 19:   The first four iterations of the Koch snowflake, showing the origin of its self-similar structure.

As Mandelbrot explained, the patterns and shapes of fractals in nature are not reproduced perfectly from one length scale to the next. There are slight variations, which is why the comparison between length scales is called *similar*, rather than *identical*. There are, however, mathematical fractals, such as the Sierpinski gasket and the Koch snowflake shown in Figure 19, in which the comparison between length scales is not just similar but exact. These objects have geometries on small length scales that are perfect copies of their overall shape.

A lot of people, including me, find fractals to be very beautiful; their beauty is probably one of the main reasons I have studied them for so long. I once heard a talk about fractals in which the speaker postulated a role for fractals in *determining* what is beautiful to us as humans.[7] He suggested that we developed our sense of beauty by being surrounded, on all sides, by fractals in the natural world and have evolved to appreciate their existence. Whether or not this explanation for the development of our sense of beauty is true, it is certainly true that fractals with their inherent regularity (the pattern of repeating self-similarity) combined with an equally fundamental irregularity are very pleasing to the eye.

Some investigators have noted that even computer-generated music can be made to sound pleasing if it is forced to take on a fractal characteristic, i.e., a degree of self-similarity. Even more interesting, a fractal structure has been found[8] to be associated with the world's greatest pieces of literature, regardless of the language they are written in.

Mysticism and the Fractal Insight

If everything from coastlines to trees to pieces of art are fractals, could it be that there is a deeper meaning to this concept? What if our most fundamental nature is a fractal one? Beyond its literal applicability to the natural world the fractal concept has struck a deep chord with many of us. It seems to illustrate, in a visual way, the philosophical principle of *microcosm* and *macrocosm*, which says that each part of the cosmos is a perfect reflection of the whole.

The idea that the nature of reality itself might be fractal is not a new one. The same idea, but with a different name, was arrived at independently by David Bohm,[9] a physicist, and Karl Pribram, a neurosurgeon. They gave the concept a different name, dubbing it the holographic view of reality.[10] This word is drawn from the concept of a hologram, which many people are familiar with from museum exhibits or as security markings on credit cards or passports. A holographic image is constructed by using lasers as the light source, but the key feature of these images is that each point within the hologram contains all the information needed to reconstruct the entire picture. The data are stored in a hologram in a self-similar way (which is what makes them so good for security codes) and, thus, reproduce the type of geometry seen in nature, i.e., fractals.

The idea that the nature of reality itself is fractal is one of the more profound of the insights I drew from the science of complex systems, but what caught — and held — my attention was the personal lesson I drew from it. If the nature of reality is fractal, did that mean the nature of my own personal relationship to it is also fractal? If this were the case, it would mean that the essence of the whole is contained within each part of myself.

One day, in the early 1990s, I had an experience that changed the direction of my work. I'd been trying to learn to meditate, but that day, as I settled onto my mat, I was overcome with a physical pain so huge it literally bowled me over. As I hit the floor, the world around me shifted.

I'd switched on some music when I settled in to meditate and as I struggled up to sit, realized I was "seeing" the music. I didn't actually see it in the normal, visual way with my physical eyes. Rather, I sensed it as a network of colored lights throbbing around me in the room. Furthermore, I seemed to have control over it. As I moved my hands, remarkably, the music changed.

The truth hit me in a flash: I was connected to this music. I was, in fact, part of this network. This was also true of everything around me — the floor I lay on, the meditation mat, the tape player, the tree outside my window. I knew I was seeing the world whole, as it really is: One great reality, with no distinction between me and you, us and them, the tree and the music. Although the experience lasted only a few moments, it was profound and like nothing I'd ever encountered before.

I've related this story chronologically as if one thing happened after another, but that's not at all the way it was: everything happened at once. It was as if time itself had no meaning — or, perhaps, as if I was somehow lifted out of time. None of that seemed strange until I tried to write it down and realized I didn't know what had happened first — or last.

A few months after I had the encounter with the network of colored lights and had managed, somehow, to completely forget it, I came upon a book in a bookstore. Its title, *Mysticism*, was a word I had never heard used in any sort of positive way. Generally, it came up in conversations with other scientists accompanied by a sneer or a rolling of the eyes, as in: "His theory is just hocus-pocus, more mysticism than science." I vaguely associated mysticism with the occult, but knew nothing about the topic; indeed, I didn't even know it *was* a topic.

The book, written by Evelyn Underhill in 1910, had a subtitle: *The Nature and Development of Spiritual Consciousness*. I leafed through several pages and was drawn to a quote by Bernard Holland: "Heaven is Nature filled with divine Life attracted by Desire."[11] Despite the old-fashioned capitalizations,

I could see what Holland was getting at. His words brought to mind an image, drawn from the pages of my lab notebooks. The image was that of an attractor and it floated in some sort of spiritual state space that this fellow Holland called Heaven, tugging and pulling on an individual Life — perhaps my life — by a force named Desire.

I read on and discovered more quotes from Mr. Holland: "In a deep sense, the desire of the Spark of Life in the Soul to return to its Original Source is part of the longing desire of the universal Life for its own heart or centre. Of this longing, the universal attraction, striving against resistance, towards a universal centre, proved to govern the phenomenal or physical world, is but the outer sheath and visible working."[12]

Holland was using gravity as a metaphor for something involving this thing-I-knew-nothing-of: mystical experience. I flipped through a few more pages. Underhill was now quoting St. Augustine (finally somebody I had heard of): "My love is my weight."

Augustine also used gravity as a metaphor. He tried to explain his experience of love for God using the attraction by a physical force, weight, as a metaphor for that love.

And, although I should have known very little of what either of these people were talking about, I knew *exactly* what they were talking about: because I had experienced this too. And, furthermore, if someone had asked me for a physical metaphor, I would have used a complex system governed by an attractor — not gravity — as a metaphor to describe the experience.

So here was my new attractor, moving me with an intense desire to learn as much as possible about a subject I had never encountered before that day in the bookstore. My scientific background was a help, in one sense: I knew how to read and dig into the literature to learn as much as possible about an unfamiliar topic. In another sense, though, it was a distinct hindrance; everything I picked up to read on the subject of mysticism generated an alarmed rationalist backlash inside my head. How can all this be true? Also, how could anybody ever *prove* a mystical experience, if it happened to them?

The fractal insight is just one example of this overlap between the scientific understanding of reality and insights about the nature of the cosmos described by the mystics. The history of astronomy is full of striking examples. Consider Giordano Bruno, for example, a 16th-century Dominican monk from Naples. He was eventually imprisoned by the pope, burned at the stake for his views, and has become a martyr of modern science for his astronomical discoveries and theories. He was also a mystic. Borchert, in his book[13] about the great mystics, explains that in Bruno's vision of the cosmos we encounter both the system of Copernicus and the infinite universe of Nicolas of Cusa.

Even more to the point here, Borchert emphasizes Bruno's use of the concept of "monads", an idea later elaborated on by Leibniz, in which "the smallest unit into which the world can be divided is, at the same time, the world as a whole."[14] In other words, Bruno (and Leibniz, too) understood the essence of reality in terms of the fractal concept: the smallest infinitesimal part of the universe *is* the universe as a whole. Here, four centuries ago, in the writings of one of science's founders about his own mystical experience, is the same idea described by physicists in our century as a holographic or fractal universe. We are all saying the same thing; the words used in each case are different, but the meaning is identical.

This view, that every part of the universe contains the entire essence of all of reality, says that existence is inherently self-similar — that reality itself is fractal. So, although Benoit Mandelbrot coined the term fractal in 1977 with the publication of his first book on the topic, and his and other contemporary works have revealed how deeply significant the idea is, the fractal *concept* is clearly not new. Bruno and Leibniz were writing about it hundreds of years ago, although without using the name coined by Mandelbrot, of course, nor the one associated with holograms, an invention of the late 20th century.

As is often the case, though, the same point can be found in ancient writings much older than those of Bruno and Leibniz. My readings in the area of mysticism soon led me East, toward Buddhist scriptures and eventually other Eastern religions, where I found a striking example of the fractal insight in a very old piece of writing from the Mahayana branch of Buddhism. A section of the Avatamsáka sutra (or scripture) called the Tathagatotpatti, or "The teaching regarding the source from which tathagatas arise", is a beautiful and moving account of the way in which the Buddha (a tathagata, or "one who has thus come") found his calling. A portion of the scripture as translated by Luis Gomez explains: "It is as if there were a sutra scroll...and on this scroll would be recorded all things, without exceptions, in this world system. This sutra scroll thus containing the world system...would be contained in a minute particle of dust. And every particle of dust in the universe would in the same way contain a copy of this sutra scroll."[15]

The scripture goes on to explain that, at a certain time, a person would appear in the world who could perceive the existence of this scroll, on which everything about the universe is written. This person would devise a means to break open the dust particle, letting out the contents of the scroll so its message could benefit all sentient beings, and the one who did this would be called the Buddha. The truth the Buddha was to reveal, then, was a truth already written in every particle of dust — and in every cell of every body. According to this ancient scripture, this truth about the fractal nature of reality is written into the very fabric of matter, the stuff of which the universe is made.

The belief, prominent in Buddhist practice, that turning inward through meditation or contemplative prayer will bring the meditator into contact with the fundamental reality of the universe is, then, a statement of belief in the universe's self-similarity. Otherwise, turning inward to contemplate oneself would be nothing more than self-absorbed navel gazing. When I first sat down to try my hand at meditation, I certainly didn't expect to come into contact with the very essence of the universe. And, yet, that is exactly what happened.

According to most meditation teachers, what is within *me* is the *part*, and that *part* contains the essence of the *whole*, so meditation will bring me into contact with that whole. And sometimes it actually does, as I believe it did for me that day when the network of light interrupted my own intense contemplation of physical and emotional pain. I was completely focused on myself, at first, and was caught off guard by the rush of knowledge that swept in to take the place of the pain.

When meditation leads to a mystical experience, the meditator is swept into a state in which knowledge of the essential nature of the whole is immediate and complete. The Buddhist tradition refers to this as achieving enlightenment or, quite simply, awakening. The idea is that enlightenment occurs when we wake up from our previous state of "unknowing", a state where we are unable to see reality as it truly is. Borchert defines a mystical experience as not just a matter of feelings or understanding. "It is a realization — with one's whole being — that all things are one, a universe, an organic whole into which the self fits."[16]

I do not know if what I experienced that day would be accepted as "enlightenment" by everybody, but it doesn't much matter to me what my experience is called. It was a profoundly frightening one, in fact, so immense that I could not hold it within my mind for any length of time. If I hadn't been compelled to write everything down, I would never have remembered that it happened. It was, however, exactly what I needed at that point in my life, and the effects have been far-reaching.

As Barbara Bradley Hagerty explains in her book[17] *Fingerprints of God*, those of us who have had these types of experiences go through a subsequent period in which our brains are "rewired" in some way, our life is irrevocably changed, resulting in a new life that is often devoted to the insights that were achieved in that moment of enlightenment or mystical experience or whatever one might choose to call it.

So I may have achieved enlightenment for one brief, shining moment, or I may have broken through into some sort of mystical understanding of life and my part in it, or I might have experienced faith in the way Paul Tillich[18] describes it: "Faith is the state of being grasped by an ultimate concern." This definition of faith is not one that made any sense to me until that day on my meditation cushion when the network of light brought me to an awareness that had always been there, even though I had never known it in any conscious way.

This type of faith is not one we can "decide" to have. It comes from either the outside or somewhere deep within — or maybe these are saying the same thing. The main point is that it grasps us without our conscious choice. I have come to think of this grasp of faith as the ultimate evidence for a Great Attractor at work in my life. When I was least expecting it, this aspect of God, the one who works on my life through my desires and interests, broke through and planted in my heart a whole new set of wishes and interests, and I have spent all the subsequent years aligning myself with these new tugs of attraction that had not been there before.

The great mystical poet, Rumi, says[19]: "O wandering Sufi, if you want to find the great treasure, don't look outside — Look inside, and seek That." As I learned so dramatically that day on my meditation cushion, the easiest way to understand the self-similar nature of reality can be summed up in a simple commandment: Know Thyself. This is a truth that has been known for centuries.

The Upanishads[20] provide another example that shows just how long this truth has been known to humanity. These classic Hindu scriptures are widely considered to be the earliest recorded accounts of mystical experience. In the Chandogya teaching of the Upanishads, a teacher explains to a student, Svetaketu, how the universe was created. This creation story clearly espouses the view that all of reality is self-similar: "In the beginning there was Existence alone — One only, without a second. He, the One, thought to himself: let me be many, let me grow forth. Thus out of himself he projected the universe; he entered into every being. All that is has its self in him alone." The teacher concludes with a widely quoted declaration to the student: "And that, Svetaketu,

That art Thou," meaning that the student — indeed, *all* of us, are the same as the One referred to in this teaching. A practitioner of Hinduism might say it this way: Atman (the human soul) *is* Brahman (the world soul).

Mystical insight is sometimes achieved as a result of a deep exploration of the self. But it can also occur spontaneously and seemingly without any attempt at self-evaluation, as apparently occurred for the poet Kathleen Raine at a young age as she sat on her soft grass-covered ledge, her secret shrine, sensing the world circling around her "day and night, wind and light, revolving round me in the sky."[21]

Mystically achieved insights can be (and, indeed, have been) dismissed as being unprovable or untestable. Sometimes, the insights achieved through mystical experience are considered peculiar to the individual mystic and reflective only of that individual's psyche rather than a legitimate insight into something fundamental about the nature of existence. It is striking, though, how similar the accounts of mystical experience are, crossing cultures and religions and thousands of years. This is, to me, evidence that these insights are not subjective but, rather, universal.

In my view, skepticism about mystical experience reflects a deep distrust of the subjective experiences of others. I suspect that this level of doubt might be overcome only when the doubter is plunged into their own mystical experience and recognizes the essence of their own otherwise baffling experience in the descriptions of others.

Even if I *had* heard directly about a mystical experience before I encountered it on my own, I doubt I would have comprehended. When I did, finally, begin to understand and, tentatively, to talk about my experience, I quickly realized that there were those who "got it" and those who didn't. Describing a mystical experience to someone who has not had one is a little like trying to describe the world as it really is to a woman from the Middle Ages. Even if she were highly educated for her time, she would probably not believe us if we said: this world is spherical, not flat as you might think; and, as if that isn't shocking enough, this little sphere you live on is but one of many in a great system of systems, a galaxy; one day, people will walk on the moon up there and send robot explorers to other planets; even more to the point, the boundaries that separate your small village from the next, your kingdom from the neighboring one, all those boundaries that seem so important to you, are temporary and will one day be gone. If we were able to show her a picture of this world she lives on, a photo of a blue orb floating peacefully in the dark blackness of

space, she would have no reason to believe us. In her time, not only was pho-
tography unknown, but the idea that one could get far enough from the Earth
to snap a picture would seem — well, out of this world.

The world as we now understand it would seem like a fiction to our woman
from the Middle Ages. She might reject our insistence that this spherical blue
world she cannot see is real. Describing a mystical experience to one who has
never achieved "breakthrough"[22] is just as difficult.

It is my experience that those who have not personally had mystical encounters with the fundamental fabric of reality are usually uninterested in reading
or hearing about such things. I don't blame those who decide, after reading or
hearing about these experiences or listening to stories about sensing a great
unity, or cosmic oneness, that it's all bunk and probably delusional. My own
personal struggle at this time was, after all, precisely this issue: was it possible
to *know* a deep truth about reality using some sort of inner seeing-faculty that
science could never prove is real?

I did, eventually, come to accept the knowledge I had received in this most
unverifiable of ways, despite my own nearly crippling skepticism. Coming
to grips with the fact that there are multiple ways of knowing, and scientific
investigation is just one way among many, formed the core of my own struggle
throughout this entire period. So, I understand the skepticism of others but
still find it sad that these most incredible facts about the very fundamental
nature of reality are so readily dismissed by so many scientists. It is as if they
have no interest in gathering information about the universe we all find ourselves part of, despite their professed devotion to discovering the truth.

The Great Fractal Attractor

A question that almost immediately comes up when I've presented to an
audience the idea of an attractor for our lives that goes through dramatic
bifurcations at times of growth is: Do each of us have our *own* attractor? And, if
we do, is our own attractor part of a bigger one? Is our own personal trajectory
a small whirlpool in a much larger state space, itself characterized by cyclic
motion around another center? The people who ask these kinds of questions
are usually struggling with understanding how the *center* of their attractor —
their *own* center — can, simultaneously, be the center of everything that is.

I believe our own personal attractor is merely the *part*, but a part that
is a perfect copy of the *whole*. Imagine a *Great Attractor* that is fractal,

composed of many smaller personal attractors, each a perfect copy of this one Great Attractor. Envision something like a spiral galaxy. In each local star system, the planets revolve around the center, their own star, but the entire star system, or solar system, rotates around the center of the galaxy. Notice that the planets in each star system rotate around *two centers* simultaneously: one center is the location of their own sun, the other is the center of the galaxy itself.

There is no contradiction here; the planet's orbit sketches a very complex pattern, but it can easily be broken down into two circles: a small circle centered on that planet's own sun, and a much larger one whose center is the hub of the galaxy. The galaxy has a fractal structure: circles on a small scale, a much bigger circle on the whole-galaxy scale. Our own galaxy is, of course, just one among billions. Do they all revolve around some unknown central point in the universe? It's possible. Perhaps the self-similarity continues *ad infinitum* — and, in essence, the center of it all will be everywhere at once.

If this were so, the urge to connect with the very center of oneself through meditation, centering prayer or contemplation might not be as self-indulgent as it is sometimes made out to be. If each individual life is a microcosm of the whole, coming to know oneself would be an excellent — and direct — way to arrive at reliable knowledge regarding the nature of everything that is. Joanna Macy, an influential interpreter of Buddhism for people in Western culture, spoke of her first experiences with Buddhist meditation as a struggle between her desire to find a "personal God" and her experience, achieved through meditation, of her own mind as a microcosm of the whole: "I knew that the nature of reality...had to be more than my mind in every respect. And since my mind has personality and intelligence and love, then it [reality] must include all that too and be not merely principle but also a personality in a much larger sense. I clearly didn't invent being a person. There is a personness writ large of which I am a small reflection."[23]

This isn't the same as making God in *our* image, though — it is merely a statement of the fact that our consciousness and other attributes of "personness", as Macy calls it, can tell us something profound about the whole of which we are a part — *if* reality is self-similar. Perhaps this is what it means to be made in God's image.

It took me a long time to come to grips with the fractal nature of my spiritual life. If the part (the individual) is *the same* as the whole, maybe not in *every* detail, but in some fundamental way, can divinity then be fundamentally different from us? Can divinity be somehow greater than us in a way that goes beyond being merely bigger? I had been taught, my whole life, not just through my religious upbringing but also through subtle messages in our culture, that it is blasphemy to make yourself out to be the same as God.

As Stephen Levine writes[24]: "Our underlying enormity frightens us." That we have, within us, the entire cosmos (an admittedly frightening idea) is very difficult for the average person to accept. Occasionally, we may be forced to acknowledge this reality, but anyone who has encountered the truth of their "inner enormity" has also felt the almost immediate urge to push this new-found terrifying fact away, as I did when I put my journal away, trying to forget all about the network of light. It apparently didn't work.

Perhaps we are a little like the spiritual seeker, Markandeya, in the Hindu story recounted by Diana Eck in her book *Encountering God from Bozeman to Banaras*. Markandeya "...was said to have roamed the whole world as it is contained within the body of Vishnu until one day he fell out of the mouth of the sleeping Lord and beheld the Lord from the outside, a sight simply too vast to comprehend."[25] Markandeya was quite fortunate, because Vishnu mercifully swallowed him again, eliminating the need to hold the overwhelming knowledge of such vastness in his own inadequate mind, returning him to a world he could grasp. It seems likely that the author of Hebrews 10:31 had a similar experience in mind when he wrote: "It is a fearful thing to fall into the hands of the living God." If the knowledge of the vastness of divinity is too much for our limited minds to bear, imagine how much more overwhelming is the sudden realization that all that vastness is contained within oneself.

Ironically, it is not just our underlying enormity that frightens us, but also its opposite. We can't help but feel small and insignificant when we think of the vastness of the universe and our vanishingly small place in it: one person, in one house, in one neighborhood on a planet in a galaxy that is, itself, only one of millions and millions of galaxies. Our life might pale into insignificance if we allow ourselves to consider this fact. Trains of thought like this make some people feel small, completely unimportant, and about as far out of touch with any "underlying enormity" as one can get. To think of our small role in the vast play of creation that is the whole universe can leave us feeling dangerously lonely and irrelevant. When Carl Sagan spoke about the billions and billions of stars in the universe, a lot of people felt only nihilistic despair and thought he was pointing out our utter insignificance — when, in reality, he was trying to communicate awe about the specialness of our selves and the inherent preciousness of our tiny world in the vast sweep of space.

Are we both utterly insignificant *and* enormous beyond belief? Yes. But then, how do we bring these two apparently opposite points of view together? The fractal concept was a great help to me in my own struggles with this seeming paradox. It can be not only an insight to help us reconcile these two opposite poles, but a source of comfort and a saving grace, as well. Here, again, this final lesson drawn from science shows us that the nature of reality is not necessarily what we've always been taught.

Through a series of crises, the turmoil of chaos, the exhilaration of mystical insight, and much reading and study, I came to see that, like the small little leaflet on a slightly larger leaf on the edge of a fern frond, each of us holds within ourselves the essence of the entire cosmos. Far from being insignificant we are each, while small and precious, a perfect copy of an even more perfect whole. When we gaze out at the stars in the night sky, it is like looking into a mirror. When, as a young girl, I wondered who I was that had sprung to life on this earth and could now look out of my two eyes on such a magnificently awesome sight as the night sky, all I had to do was consider those stars.

There, I was seeing myself. And all of you can see yourselves, too: little lights twinkling in the distance, perfect reflections of the lights shining in our homes on this planet. Far from being small and insignificant, we are, each of us, as enormous and wonderful as the whole night sky.

Endnotes

1. Jonathan Star and Shahram Shiva, translators, *A Garden Beyond Paradise: The Mystical Poetry of Rumi*, Bantam Books, New York (1992), p. 53. Permission from Jonathan Star to reprint this short poem is gratefully acknowledged.
2. Otto E. Rossler, "An Equation for Continuous Chaos," *Physics Letters*, Vol. 57A, No. 5, pp. 397–398 (1976).
3. Benoit B. Mandelbrot, *The Fractal Geometry of Nature*, W.H. Freeman and Co., San Francisco (1983).
4. Mandelbrot, p. 4.
5. For more details on the process of calculating fractal dimensions, see: Harold M. Hastings and George Sugihara, *Fractals: A User's Guide for the Natural Sciences*, Oxford University Press, Oxford (1993).
6. One especially nice book that explores fractals in nature using many illustrations is Michael Barnsley, *Fractals Everywhere*, Academic Press, New York (1988).
7. The speaker was Richard Voss. See the following paper for a more detailed description of his research: Richard Voss and John Clarke, "1/f Noise in Music and Speech," *Nature*, Vol. 258, pp. 317–318 (1975).

8. A recent study can be found here: S. Drożdż et al., "Quantifying origin and character of long-range correlations in narrative texts," *Information Sciences*, Vol. 331, pp. 32–44 (2016)

9. David Bohm and F. David Peat, *Science, Order and Creativity*, Bantam Books, New York (1987).

10. This idea has gained some traction recently in cosmological physics due to new discoveries. See, for example: N. Afshordi et al., "From Planck data to Planck era: Observational tests of holographic cosmology," *Physical Review Letters*, Vol. 118, p. 041301 (2017).

11. This excerpt from Evelyn Underhill, *Mysticism: The Nature and Development of Spiritual Consciousness*, 12th Edition, Oneworld Publications © 1993 is reproduced with permission of the Licensor through PLSclear.

12. Underhill, p. 117; reproduced with permission of the Licensor through PLSclear.

13. Bruno Borchert, *Mysticism: Its History and Challenge*, Samuel Weiser, Inc., Maine (1994).

14. Borchert, p. 264.

15. L.O. Gomez, "The Whole Universe as a Sutra," in *Buddhism in Practice*, D.S. Lopez, ed., Princeton University Press, Princeton, New Jersey (1995), pp. 107–112; permission from Princeton University Press to reproduce this excerpt is gratefully acknowledged.

16. Borchert, p. 11.

17. Barbara Bradley Haggerty, *Fingerprints of God: The Search for the Science of Spirituality*, Riverhead Books, New York (2009).

18. Paul Tillich, *Dynamics of Faith*, Harper & Row, New York (1957).

19. Jonathan Star and Shahram Shiva, translators, *A Garden Beyond Paradise: The Mystical Poetry of Rumi*, Bantam Books, New York (1992). Permission from Jonathan Star to reprint this and other short poems is gratefully acknowledged.

20. For an English translation, I recommend; S. Prabhavananda and F. Manchester, *The Upanishads*, Penguin Putnam, New York (1948).

21. Kathleen Raine, *Farewell Happy Fields*, Hamish Hamilton, London (1974).

22. It has been suggested that the medieval mystic Meister Eckhart (1260–1328) invented the word breakthrough (*Durchbruch* in the original German) to describe the key event in a mystical encounter. See, for example, Matthew Fox, *Breakthrough: Meister Eckhart's Creation Spirituality in New Translation*, Doubleday, New York (1980).

23. Interview with Joanna Macy quoted in: Anne Bancroft, *Weavers of Wisdom: Women Mystics of the Twentieth Century*, Penguin Books, London (1989), p. 9.

24. Stephen Levine, *Guided Meditations, Explorations, and Healings*, Doubleday, New York (1991).

25. Diana L. Eck, *Encountering God: A Spiritual Journey from Bozeman to Banaras*, Beacon Press, Boston (1993), p. 78.

Insight 7   Emergence

> *This vast similitude spans them, and always has*
> *spann'd,*
> *And shall forever span them and compactly hold*
> *and enclose them.*

Walt Whitman[1]
"Sea Drift"

When a large number of objects (particles, molecules, people, and so on) interact, behaviors or properties can emerge that are not possible for each individual in the group. Emergence is the process that gives rise to these collective properties and behaviors.

When the collective in question is a complex adaptive system of the type we've considered in this book, the emergent behavior can be quite elaborate. Patterns can form or oscillations can arise — all types of self-organized behavior are included in the concept of emergence.

Even systems that are not adaptive have emergent properties and behaviors, however. Consider water, for example. A single water molecule cannot be wet, or as Niels Bohr, one of the founders of quantum theory, put it, "Analyzing hydrogen and oxygen individually will not reveal the wetness of water."[2]

Wetness is a property that emerges from the interactions of water molecules. For something to be wet, it has to have viscosity and surface tension, among other things, and neither of these concepts applies to a single molecule.

Similarly, a single water molecule cannot be hot or cold — temperature is another emergent property that arises from the collective behavior of a large number of molecules.

Temperature is a measure of the average energy generated by movements of atoms or molecules in an object or substance. The absolute temperature T is introduced in a statistical mechanics derivation that connects the microscopic world of atoms and molecules, where T has no meaning, to the macroscopic world. It comes into the derivation as a scaling factor, usually denoted kT, where the quantity k is known as the Boltzmann constant.[3]

If the system in question is adaptive, though, properties and behaviors that are much more complex can emerge. As we have seen in previous chapters, these complex adaptive systems can self-organize to produce patterns, oscillations, chaos, flocking, swarming — a whole host of elaborate emergent phenomena.

The Whole is not just More — It's Different

The term "emergence" can be traced back to the dawn of systems theory, which is where the concept of the whole being more than the sum of its parts originates. Systems theory has been around for a while, with scholarly articles dating to the 1950s. It was not until 1972, however, when Philip Anderson wrote an article in the journal *Science* entitled "More is Different" that scientists really began to grapple with the fact that complex systems are qualitatively different from the isolated parts which collectively make them up. In this seminal article, Anderson wrote: "The ability to reduce everything to simple fundamental laws does not imply the ability to start from those laws and reconstruct the universe...At each level of complexity entirely new properties appear. Psychology is not applied biology, nor is biology applied chemistry. We can now see that the whole becomes not merely more, but very different from the sum of its parts."[4]

This idea — that the whole is not just *more* than the sum of its parts, but also *different from* those parts of which it is comprised — is what makes the concept of emergence so compelling. The way the parts of the system interact is key. We can't understand the mechanism that leads to emergence without considering relationships between the individual units in the system.

In addition, the behavior of the system as a whole can *feed back on* the parts, changing the behaviors of those individual parts and possibly changing the way they interact with each other. This can lead to new emergent behaviors at the level of the whole that, in turn, affect the behaviors of the individual parts, which go on to affect the whole system again — *ad infinitum*, or at least until the collective converges on a fixed and final state.

Some have suggested that thinking itself is an emergent property of the brain. Thought does, after all, have the qualifications for an emergent property: it is a behavior that has no meaning at the level of the parts. Individual neurons do not think. Also, actions of the whole system feed back on the individual parts, so while our neurons don't think, we do, which means that our thoughts affect our neurons, as well as the synapses or connections between them. This is, in fact, how we learn. It remains an open question whether consciousness, much less the thought process, is an example of emergence, but it's an intriguing possibility.

The influence of the system on the parts can make emergence seem more than a little scary to those who first hear about it. If you are one of the parts and society is the whole, the idea of the collective system imposing its will can be quite frightening. And yet, it is obviously true that society feeds back on each individual. The social system of which we are a part affects every person in it. We are all products of our culture.

Is this feedback effect something that I, as an individual, have any choice about? Or am I no different from one of those neurons in my brain that must change when the system that is my brain learns something new? We will consider these and similar questions about the deeper meaning of emergence in a later section of this chapter.

There is still a lot we don't understand about emergence, but one thing we do know: it is real, and the world is full of examples. Consider just a few: traffic, financial markets, epidemics, and — last, but certainly not least — our brains.

Traffic in a Shared Space

Traffic is one surprising area where the understanding and appreciation of emergent behavior has had an impact on design and "control" systems. I put the word "control" in quotes here, because these systems don't actually *control* traffic. Instead, they allow pedestrians and the drivers of cars, bicycles and trucks to self-organize into traffic flow patterns that are safer than the alternative.

These so-called shared space[5] traffic design systems were pioneered by traffic engineer Hans Monderman in the Netherlands. The approach is based upon the idea of removing most, if not all, traffic signs and other traditional means of traffic control from an intersection. This forces people — drivers and pedestrians alike — to be more alert to those around them. Eye contact is key and has been shown to both increase safety and reduce congestion. Here is where interactions between the system's parts (the pedestrians and drivers) come in. By making eye contact with nearby neighbors, each person engages in a subtle act of communication: "Which way are you going next?" And: "I am going this way." This information exchange is crucial to establishing a self-organized flow of traffic.

An intersection in the town of Haren in the Netherlands saw accidents drop by 95%, from an average of 200 per year to about 10 per year, after the intersection was redesigned using the shared space concept.[6] This success led to shared space ideas spreading across Europe in the early 2000s. In 2003, the European Union launched a research project to study the idea and see if it could be implemented elsewhere. Traffic engineers began to flock to Drachten, a town in the Netherlands where Monderman had set up a demonstration project.[7] Soon, shared space streets were popping up all across Europe.

The shared space philosophy is very different from traditional traffic control methods that are based on rules and signals. In a shared space situation, traffic

Figure 20:   A shared space intersection on Sonnenplatz in Graz, Austria

flow patterns emerge in the same way that flow patterns emerge in flocks of birds or schools of fish. When birds flock or fish school, no one individual is in charge, yet the group moves in an orderly fashion. The key is interaction: each bird in a flock senses the position and direction of motion of its neighbors and adjusts its own flight to match. The same thing happens in shared space traffic systems: individuals sense, through eye contact, the position and direction of motion of the people around them and adjust accordingly. What emerges is safe, smooth traffic flow.

Financial Markets Sometimes Behave like Sand Piles

Financial crashes provide a particularly dramatic example of an emergent phenomenon that is more catastrophic in nature. This type of emergence is known as self-organized criticality.

The financial market is a complex, adaptive system that often seems to be stable, until the rare catastrophic event of a market crash occurs. Physicists and economists have begun, in recent years, to develop market theories[8] that use a complex systems approach to understand financial meltdowns like those that occurred in 1929, 1987 and, more recently, 2008.

The global financial market is a complex adaptive system in the true sense of the word. It consists of many parts — individuals who buy and sell shares in companies. These people, the parts in the system, are interconnected by a network of prices and buying and selling mechanisms. Prices fluctuate over time, sometimes changing rapidly, and market agents make their decisions to buy or sell in this dynamic environment.

Electronic trading and communication systems have greatly increased the speed of interaction between the agents in the market. Some buy-and-sell orders are even made automatically using software triggered when certain market conditions cross a predefined threshold. All of these lead to the possibility that catastrophic market declines can happen over a period of mere days or even hours.

The market is a complex system because it is comprised of multiple, nested feedback loops. Some of these loops are positive, meaning changes are reinforced and amplified, whereas others are negative — damping out any trends that might start to set in. A crash will happen when a positive feedback loop in this system takes over and damping forces are not enough to stop it. In simple terms, the market will crash when agents all place sell orders simultaneously or in rapid succession, driving prices down, which leads to more sell orders and, eventually, a catastrophic plunge in market values.

As Didier Sornette, a professor of both Physics and Finance in Zurich, has pointed out,[9] it is interesting that the traders on the market do not generally know each other or communicate directly. Their interactions occur through the market itself. Each agent notices what the others are doing (buying or selling particular stocks) and makes decisions based on their observations. Sometimes the agents agree with one another and a coordinated sell-off occurs, but most of the time they disagree and balance each other out. As Sornette explains, a crash happens when the system becomes too orderly — when everyone has the same opinion, to sell. When enough agents have different opinions, disorder reigns — and the market remains stable.

Some theorists have suggested that the dynamics exhibited by financial markets is similar to that of a simple pile of sand. Both, they say, move naturally toward an unstable situation known as self-organized criticality[10] in which a tiny change, akin to the straw added to a camel's back, can have a catastrophic outcome. Self-organized criticality is, in fact, an emergent phenomenon, since it occurs only at the level of the collective.

Despite the fact that self-organized criticality leads to what we might call catastrophes — market crashes, earthquakes, and wildfires — sometimes the catastrophic change is, in the end, for the best, so there should be no judgment placed on the use of the word "catastrophe" here. Political crises and revolutions have also been treated as examples of the type of emergence characterized by self-organized criticality. What might be a negative event for some could be quite positive for others.

To understand how self-organized criticality leads to such dramatic moments of change, imagine a sand pile fed by a slow stream of individual sand grains from above. The grains tumble slowly down the sides of the pile, coming to rest at various points along the slope. The pile gradually grows as sand continues to dribble on the top until, at some critical point, a bit too much sand collects in one place and a small avalanche occurs. A rush of sand tumbles to the bottom, returning the pile as a whole to a relatively stable state where more sand can be dribbled on the top.

Sand exhibits this behavior because each grain is rough and a little bit sticky, so it tends to slightly adhere to its neighbors in the pile. A pile of frictionless marbles would not exhibit self-organized criticality. In fact, marbles would not form a pile in the first place, so no opportunity for avalanches would ever occur in a system of frictionless marbles.

Small avalanches can occur randomly at any location on the pile when the system reaches a critical size. An extra grain of sand added anywhere can cause an avalanche in the same way that a trader somewhere in the market can initiate one more sell order that is one order too many — and a catastrophic sell-off ensues. The trigger can occur any place and for any reason. The market has a systemic instability that individuals in it cannot control.

The common feature in systems that are governed by self-organized criticality is that they are driven by a persistent external force. In the case of the sand pile, the external force is the stream of sand entering at the top. For the market, it is the influx of new money from investors. For a wildfire or forest fire, it is the accumulation of dry fuel over time. And for earthquakes, the external driving force is the constant pressure from colliding tectonic plates pushing on fault lines that will, at some point, slip. What emerges, in all these examples, is catastrophic change.

When an Outbreak Becomes an Epidemic

As I write this, the world is in the grips of the deadly COVID-19 pandemic. The first vaccines are just now becoming available, so we still have many months to go before the pandemic will be controlled, much less eradicated. Throughout these months as the disease has spread, epidemiologists have used models to predict changes in infections. Public officials have relied on both current data and predictions made from simulations with these models to impose various rules and strategies to try and control the spread.

The pandemic started with a single outbreak of the SARS CoV-2 virus in Wuhan, China, in late 2019 before exploding, first into a localized epidemic, and soon thereafter into a global pandemic. To understand how an outbreak blows up to become an out-of-control epidemic or pandemic, epidemiologists use models where individuals in the population are divided into groups. One type of model that has been known for almost a hundred years is the SIR model,[11] where S stands for the number of susceptible individuals, I the number of infected ones, and R the number of people removed from the system — either recovered, quarantined or isolated, or deceased.

The dynamics of a model like this depends on a number of factors, including an important one known as the reproduction number, which is the number of people one infected person can spread the disease to. If this number is larger than one, the disease has the ability to become an epidemic. Whether or not the spark of an outbreak flares up into a full-blown epidemic or pandemic depends on the details of the dynamic model. It depends especially on the rate at which infected individuals are removed from the system — preferably through self-isolation or recovery. This is, of course, where various mitigation strategies — quarantine, lockdowns, masks, etc. — can affect the course of the spread.

Many years ago, my colleague, Lars Folke Olsen, along with his collaborator, William Schaffer, used a complex systems approach to study historical data for childhood infectious diseases prior to the development and introduction of vaccines for them. They analyzed data for measles, mumps, rubella, chicken pox, and other diseases, all of which exhibit noisy annual oscillation cycles. Olsen and Schaffer were able to show[12] that many of these "noisy" oscillations were actually chaos of the type we considered in Chapter 5.

Epidemiologists have been slow to adapt the type of analysis techniques that Olsen and Schaffer used, but they do seem to appreciate that epidemiology would benefit from this approach. In an editorial entitled "Complexity, Simplicity and Epidemiology" in the January 2006 issue of the *International Journal of Epidemiology*, Neil Pearce and Franco Merletti note, "There are very few examples of the use of complexity theory in epidemiology, but there are many examples of epidemiological problems for which complexity theory is relevant."[13]

Some recent studies have used network theory, one tool in the complex systems analysis toolbox, to understand how outbreaks spread when people can move rapidly around the globe, traveling by air. In 2007, my former colleague, Alessandro Vespignani, and his co-workers used a computational network model to understand[14] the global spread of SARS, the severe acute respiratory syndrome that was caused by a virus similar to the one that causes COVID-19. This type of complex systems approach greatly increases the forecasting ability of models for investigators seeking to understand the global spread of an outbreak.

Dynamical Diseases and the Brain

In the mid-1990s, I received an inquiry from Dr. Robert Worth, a neurosurgeon at the medical school on our campus. He had been reading about chaos theory, he said, and wanted to ask some questions since he thought it might have something to do with epilepsy. His patients, young people with a form of epilepsy that had become resistant to medication, underwent surgery for their illness, but no one knew why the surgery worked, nor why it failed in more than 10% of the cases.

When Bob called with his inquiry, I initially assumed he was interested in determining if a seizure was an example of chaos — but it turns out that the opposite is true. A seizure is a highly synchronized, orderly state of brain activity. Chaos, he said, is a sign of healthy brain function. When I realized what Bob was saying, I knew I had to learn more about this illness and why it was that he thought theories about chaos would help.

We initiated a collaboration, which was hampered by the two of us having very different scientific backgrounds and completely different vocabularies. He tutored me on basic neurophysiology and I tutored him on the mathematical and physical theories of complex systems. A graduate student, Brent Speelman, joined the effort and the three of us began to develop a computer model[15] that we hoped would capture the key features of brain activity that led to seizures. We also hoped our model would lead to new insights and treatments for epilepsy and, possibly, other so-called dynamical diseases of the brain.

Dynamical diseases are those associated with major changes in the dynamics of various bodily functions.[16] They arise in many systems, including the heart and other organs, but are especially striking when brain activity is involved. Epilepsy is considered to be a dynamical disease, but other disorders are as well, such as Parkinson's disease. Dynamical diseases of the brain usually arise at the level of populations of neurons, not the level of single cells (see Figure 21).

The type of seizure we were concerned with is known as a *complex partial seizure*. These seizures are thought to originate from an area of abnormal tissue in the brain, usually in a region known as the hippocampus, a portion of the temporal lobe. This *focal abnormality*, as it's called, could have originated through a blow to the head, low oxygen during the birth process, or other causes. As Bob explained to us, patients with this type of injury do not begin having seizures right away. Abnormal firing patterns in the focal region exist, but they can be damped out or dissipate without any adverse effects.

Over time, however, these abnormal patterns become more fixed and begin to spread, recruiting healthy neurons in the surrounding tissue until they urge the entire brain to fire in synchrony. It is, Bob said, as if the brain learns to have a seizure, adding synapses, new connections, and pruning others out, until the focal abnormality "teaches" the surrounding neurons how to fire at the same time.

Levels of Organization

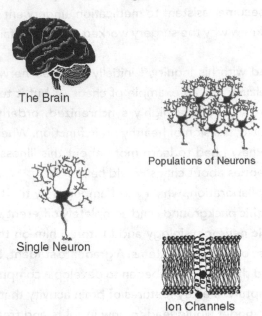

The Brain

Populations of Neurons

Single Neuron

Ion Channels

Figure 21: Thought, memory and even consciousness arise at multiple levels of organization in the brain. Ion channels and neurotransmitters govern our brain's behavior at the lowest, molecular level. Neurons and populations of them govern it at higher levels.

A seizure has been likened to a crowd in a football stadium seized with the same emotion when a player on the field scores a goal. The player's excitement is transmitted to the crowd, who previously had been talking to each other in random, disorganized fashion. When the goal happens, the attention of every-one in the stadium is diverted to the action on the field, and the crowd rises up as one unit with a big roar. This is very much like what happens in a seizure. The focal abnormality sends out a signal and the surrounding neurons in the brain all begin to fire at the same time.

The model that we developed involved two types of brain cells, both neu-rons (see Figure 22). One type is known as *excitatory* neurons (more specifically *pyramidal* cells, so-called because of their shape); the other type as *inhibitory interneurons*. The excitatory neurons form a positive feedback loop, while the inhibitory ones are involved in negative feedback.

Each cell type was fed by an input current from another part of the brain known as the *perforant pathway*. The two cell types were connected in a small network (see Figure 23) referred to as a *subnetwork*, since it was, in turn, con-nected to the full neural network of the brain. In our model, we connected

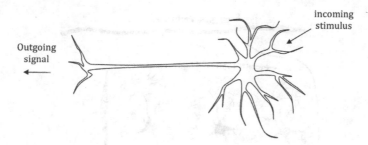

Figure 22: Schematic drawing of a neuron, showing the branched structures that receive and transmit electrical signals as well as the long axon along which the stimulus pulse travels.

these small subnetworks using a mechanism involving the diffusion of potassium, which neurologists believed was involved with seizure propagation through healthy brain tissue.

We used physical values for this model drawn from experimental studies of brain tissue and were, thus, able to carry out simulations and calculations with the model that could be compared directly to clinical data. We found that lower levels of inhibition led to seizure-like activity in our model. Another way to say this is that positive feedback effects dominate during a seizure. Negative feedback, through inhibition, cannot stop the surge of activity from spreading. We were excited by this result, since it agreed with clinical studies that suggested a reduction in the release of an inhibitory neurotransmitter known as GABA could cause seizures.

One of the key features of our model was the mechanism that connected small subnetwork regions to each other. Our model, a mathematical description, was consistent with at least three different physical mechanisms. One of these, and the most intriguing one, involved electrical communication through a third type of brain cell known as a *glial cell*. These cells were once thought to be merely the "garbage collectors" of the brain, soaking up waste products from cellular function and moving them out into the blood stream. Now, however, they are known to be intimately involved with transmitting information around the brain. Our results suggested that when transmission through the glial network became too fast, a seizure could spread and engulf the entire brain.

Our simulations showed that a donut-shaped attractor (known, mathematically, as a *torus*) governed the behavior of our model. We found simple, periodic behavior for low degrees of inhibition. For higher degrees of inhibition, the model exhibited a cascade of period-doubling bifurcations as the torus attractor broke up into a fractal object. The end result was, ultimately, chaos. Note,

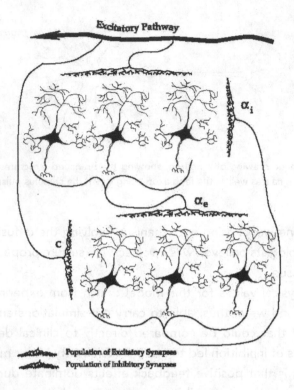

Figure 23: Schematic illustration of the subnetwork model used to study epileptic seizures. The top neurons form synapses labeled α_e (for excitatory) with the lower group of synapses. They, in turn, inhibit the top row through synapses labeled α_i.

however, that chaos is associated with a *non-seizure* state. Chaos is the way healthy, normal brains function, whereas periodic behavior is observed when the brain activity becomes synchronized — in other words, during a seizure.

When inhibition was completely absent from our model, however, the simulation went to a fixed point — stasis. In other words, a situation with no inhibition (no negative feedback) leads to a state resembling death. Too much inhibition leads to orderly behavior: a seizure. The middle ground, a balance of positive and negative feedback, leads to disordered behavior. This is the healthy state. Chaos, it turns out, is a sign of health when it comes to our brain.

Social Emergence

Humans are social creatures and we have found myriad ways as a species to self-organize. Fads and fashions spread literally and figuratively like wildfire through our populations. The outcome can be good, bad or neither — the point is that the mechanism by which fads spread is the same as the one that governs the spread of ideas.[17]

Innovation and new ideas arise and move through the culture by a variety of means, including some slower means of communication such as books and education. As a species, we have developed an impressive number of examples of collective behavior — science, religion, medicine, manufacturing, transportation and communication infrastructure, not to mention art and music — but none of this would ever have existed without the ability to self-organize and adapt that is so characteristic of our species.

And yet change, when it occurs, is often messy, disorganized or even, at times, violent. Calls for reform almost always bubble up from the street — the "grassroots", as people refer to it — rather than being imposed from the top down. This is clearly emergence — behavior that exists at the level of the collective but is not possible for a single unit, i.e., individual, within it.

An interesting example is suggested by Phyllis Tickle in her book *The Great Emergence*. "The Church," Tickle writes, "is a self-organizing system of relations, symmetrical or otherwise, between innumerable member-parts that themselves form subsets of relations within their smaller networks, etc. etc. in interlacing levels of complexity."[18]

In this book, Tickle argues that the Christian church has gone through transformational changes every 500 years or so, and is going through another now. These cycles are tied to changes in communication patterns. 500 years ago (or thereabouts) Martin Luther nailed his treatise to a church door and initiated the Great Reformation. 500 years before that, the church split into the East and the West: Eastern and Greek Orthodox on the one hand, Roman Catholicism on the other.

Going back 500 more years (roughly), we arrive at the fall of the Roman Empire and the ushering in of the dark ages. And 500 before that was the time of the historical Jesus, ushering in a transformation so great that even the way we number our years changed. This surprising periodicity continues backward into Judeo-Christian history, since 500 years before Christ we have the fall of the temple in Jerusalem and 500 years before that was the reign of King David.

Tickle argues that without Gutenberg's new printing press, Luther's ideas would not have spread and the break between the Protestant church and the Catholic church would not have happened. The Church is changing again, she argues, because we now have the Internet and social media.

These technologies make leaderless, yet organized, revolutions possible. Consider, for example, the spring of 2011, which was referred to that year as the "Arab Spring", since societal unrest seemed to promise a new day of self-determination and freedom. The revolution popped up in multiple places as crowds of demonstrators gathered in town squares across the Arab world, protesting the actions and policies of the dictators that ruled their countries. Tahrir Square in Cairo, Egypt, quickly became the epicenter of this revolt.

The revolution was different in one way from previous revolutions since the organizers reached like-minded individuals through technological means, including Twitter, Facebook and other social media. An uprising in Tehran, Iran, two years earlier had started the same way, with protestors using Twitter to organize their demonstrations. The authorities tried to quash the revolt by cutting off access to the Internet and trying to find and arrest people by looking at their location on Twitter.

This action by the government to suppress the uprising led to a remarkable event: sympathizers around the world began changing their own locations in their Twitter bio to Tehran, even if they were actually in London, Washington, or wherever. There was also a concerted effort by people outside Iran to redirect Twitter messaging to hidden servers in secret places so the protestors in Tehran could continue to communicate with each other and evade detection and possible arrest.

Two years later, the uprising in Cairo began in a similar way, with social media bringing people together in Tahrir Square to demand that President Mubarak resign. The protestors in Egypt did not seem to be inspired by a particular ideology but, rather, by a desire for freedom from oppression. There was no single charismatic leader, a fact that seems to have bewildered many in the media and opinion sphere.

It was this characteristic — a leaderless revolution — that led some of us to wonder if we were witnessing a real-life example of self-organization in the human sphere. We heard remarkable reports of behavior that certainly sounded like self-organization: people divided up tasks, some guarding entrances to the area. Others prepared food, distributed water, provided medical help. In other words, the people governed themselves.

This should not be surprising, since people have been doing it for millennia. Civilization would not exist were it not for our inherent abilities to organize ourselves into functioning societies. It is not possible, though, to predict that a revolution that begins in a self-organized way will always lead to greater democracy.[19] In fact, the opposite has often been true, since there will always be those who seek to grab power and they may not use it for the good of all.

The Arab Spring was just one example of many self-organized uprisings in human history. The United States has been the scene of several more over the last few years: the Women's March, the March for Science, protests and legal actions against immigration policies and, more recently, the Black Lives Matter movement. All of these were, like the 2011 uprisings in the Arab world, initiated and accelerated by high-speed communications through text and social media.

It is a lot easier to bring people together with cell phones than it is by riding a horse through the streets shouting, "The British are coming!" as Paul Revere did in 1775 to launch the American Revolution — and, yet, the end result was much the same. The people self-organized and what emerged has changed the world.

Endnotes

1. Walt Whitman, "Sea-Drift: On the Beach at Night Alone," in *Leaves of Grass*, 1882 Edition: https://en.wikisource.org/wiki/Leaves_of_Grass_(1882). This poem is in the public domain.
2. Niels Bohr quoted in Richard H. Jones, *Science and Mysticism: A Comparative Study*, Bucknell University Press, Lewisburg (1986), p. 38.
3. Donald A. McQuarrie, *Statistical Mechanics*, University Science Books, California (1984).
4. Philip Anderson, "More is Different," *Science*, Vol. 177, p. 393 (1972).
5. Ben Hamilton-Baillie, "Shared Space: Reconciling, People, Places and Traffic," *Built Environment*, Vol. 34, No. 2, pp. 161–181.
6. Craig Whitlock, "A Green Light for Common Sense," *The Washington Post*, Dec. 24, 2007.
7. A short video of traffic in the Drachten intersection is available on YouTube. See: https://www.youtube.com/watch?v=tye8zJr7pZ0.
8. (a) H.E. Stanley, L.A.N. Amaral, S.V. Buldyrev, P. Gopikrishnan, V. Pierou, and M. Salinger, "Self-organized complexity in economics and finance," *Proceedings of the National Academy of Sciences*, Vol. 99, pp. 2561–2565 (2002).
   (b) Leopoldo Sánchez-Cantú, Carlos Arturo Soto-Campos, and Andriy Kryvko, "Evidence of self-organization in time series of capital markets," Statistical Finance. *arXiv*:1604.03996 (2016).

(c) S.C. Nicols and D.J.T. Sumpter, "A dynamical approach to stock market fluctuations," *International Journal of Bifurcation and Chaos*, Vol. 21, No. 12, pp. 3557–3564 (2011).

(d) P.W. Anderson, J.K. Arrow, and D. Pines, eds., *The Economy as an Evolving Complex System*, CRC Press, Taylor & Francis Group, Florida (1988).

9. Didier Sornette, "Predictability of catastrophic events: material rupture, earthquakes, turbulence, market crashes and human birth," *arXiv:cond-mat*/0107173v1 (2001).

10. (a) P. Bak, C. Tang and K. Wiesenfeld, "Self-organized criticality," *Physical Review Letters A*, Vol. 38, No. 1, pp. 363–374 (1988).

(b) A. Chessa, H.E. Stanley, A. Vespignani, S. Zapperi, "Universality in Sandpiles," *Physical Review E*, Vol. 59, No. 1 (1999).

(c) J.A. Scheinkman and M. Woodford, "Self-organized criticality and economic fluctuations," *The American Economic Review*, Vol. 84, No. 2, pp. 417–421 (1994).

11. W.O. Kermack and A.G. McKendrick, "A contribution to the mathematical theory of epidemics," *Proceedings of the Royal Society A*, Vol. 115, pp. 700–721 (1927).

12. (a) L.F. Olsen, G.L. Truty and W.M. Schaffer, "Oscillations and chaos in epidemics: a nonlinear dynamic study of six childhood diseases in Cophenhagen, Denmark," *Theoretical Population Biology*, Vol. 33, pp. 344–370 (1988).

(b) L.F. Olsen and W.M. Schaffer, "Chaos vs. noisy periodicity: alternative hypotheses for childhood epidemics," *Science*, Vol. 249, pp. 499–504 (1990).

13. Neal Pearce and Franco Merletti, "Complexity, simplicity, and epidemiology," *International Journal of Epidemiology*, Vol. 35, pp. 5105–519 (2006).

14. V. Colizza, A. Barrat, M. Barthélemy, and A. Vespignani, "Predictability and epidemic pathways in global outbreaks of infectious diseases: the SARS case study," *BMC Medicine*, Vol. 5, No. 34 (2007).

15. B. Speelman, R. Larter and R.M. Worth, "A Coupled ODE Lattice Model for the Simulation of Epileptic Seizures," *Chaos*, Vol. 9, pp. 795–804 (1999).

16. J. Belair, L. Glass, U. an der Heiden *et al.*, "Dynamical disease: identification, temporal aspects and treatment strategies of human illness," *Chaos*, Vol. 5, pp. 1–7 (1995).

17. Malcolm Gladwell discusses this idea in his book *The Tipping Point*, Back Bay Books, New York (2002).

18. Phyllis Tickle, *The Great Emergence: How Christianity is Changing and Why*, Baker Books, Michigan (2012), p. 152.

19. The *New England Complex Systems Institute* has considered this question through a detailed analysis of the events of the Arab Spring. See https://necsi.ed/sixyear-report-on-the-arab-spring.

Conclusion

In finding our path through life, we encounter difficulties and sometimes come upon moments of transformational change. This is inevitable. As we have seen in the preceding pages, however, what can seem like the end of the world might very well be the beginning of a new life. This type of change can happen to us as individuals, members of a group or even humanity as a whole. We would be well-served to draw insights from the natural world in finding ways to navigate turbulent times.

As I am completing this manuscript, the human race is caught in the clutches of a tenacious virus, SARS CoV-2. This virus has swept through country after country, spreading from one individual to another, infecting millions, and crippling economies everywhere. As of this writing, over two million people have died worldwide from COVID-19.

When this all started, the future seemed bleak at best. The devastation has, in fact, been severe. Our species, though, reacted in the way we always have: by adapting, finding ways to change behaviors that would thwart this tenacious enemy, and getting to work in the lab to find cures and, eventually, vaccines that might very well put an end to this pandemic. The process has been far from smooth, but we have made definite progress.

Human society is a complex adaptive system and this past year has proven that in more ways than one. We have been caught in seemingly endless cycles where one day felt exactly like the next, only to be followed by periods of chaos where many worried that the whole system would come crashing down. Civilization is, after all, a fragile thing and it only makes sense to wonder if we are resilient enough to withstand an assault like the one this virus has mounted on us. Is there an attractor that stabilizes human society? Are we merely going through a bifurcation? Will our world undergo a dramatic change now, coming out the other side in a new form, with a new attractor to govern our future trajectory?

I, of course, don't know the answers to any of these questions, but it strikes me that the pandemic is just one among many threats facing our world. Another that is truly an existential threat is global warming. Governments have taken to using what seems to me to be a euphemism for this. Climate change describes this warming, but when we say "change", it implies we don't know which way the change is taking us. We *do* know: the concentration of carbon dioxide in the atmosphere is going up, the average global temperature is rising, glaciers are melting and sea levels will rise. These are the facts.

It remains to be seen if our species is resilient enough to adapt to this looming crisis. If humanity can overcome one disease after another, extend the average lifespan, send members of our species to the moon and exploratory robot ships to Mars and Pluto, develop communication and information technologies that link our world into one high-speed whole — perhaps we can overcome the crisis of global warming and climate change and will, one day, look back on this moment as a bifurcation when our old world gave way to a brand new one.

Remember: what seems like the end of the world to the caterpillar is, to the butterfly, the beginning of a whole new life.

Glossary

attractor A graph of a system's trajectory toward which that system evolves for a wide variety of starting conditions.

Arab Spring The spring of 2011, characterized by uprisings across the Middle East, including Tahrir Square In Cairo, Egypt.

bifurcation A sudden qualitative change in an attractor's character that occurs as the result of a small change in a key parameter.

Belousov-Zhabotinsky reaction A mix of several inorganic chemical species, usually including the indicator ferroin that exists in both red and blue forms. The reaction is named for its discoverers, Boris Belousov and Anatoly Zhabotinsky.

Bruno, Giordano A Dominican friar, mathematician and cosmologist, born in 1548, who is known for his extension of the Copernican model. Tried by the Roman authorities for heresy, he was burned at the stake in 1600.

butterfly effect A short-hand phrase for the phenomenon of sensitivity to initial conditions, in which a small change in starting conditions, such as the flap of a butterfly's wings, can lead to a much greater change later.

cardiac pacemaker cell Specialized cells in the sinoatrial node of the heart which create the rhythmic electrical impulses that cause contraction and pumping of blood by the heart muscle.

carrying capacity The maximum population size of a species that can be sustained in a given ecological environment.

catalyst A chemical substance that increases the rate of a chemical reaction.

cell The smallest unit of life, generally consisting of a cell membrane surrounding a liquid-like substance known as protoplasm that contains a variety of tiny cellular structures such as the cell nucleus, chromosomes, mitochondria, and others.

chaos In the context of this book, chaos is deterministic, aperiodic motion in state space, characterized by sensitive dependence on initial conditions and generally associated with a fractal attractor.

chemical reaction A process that converts one set of chemical substances into a different set. Chemical reactions, as opposed to physical processes, always involve the making and breaking of chemical bonds between atoms.

Choristoneura freemani A species of moth whose larva, known colloquially as the western spruce budworm, wreaks destruction in the pine forests west of the Mississippi.

Cicada Insects with large eyes and nearly transparent wings that live near trees and lay their eggs under the bark. Periodical cicadas spend many years, 13 or 17, underground as nymphs.

classical mechanics A system of equations and theories that describe the motion of macroscopic objects such as billiard balls, machines or even planetary objects in terms of their position, velocity and acceleration.

collective Another word for system.

complex system A collection of many individual parts that interact in such a way that the system becomes adaptive and exhibits emergent properties.

complexity theory The study of complex systems.

creationism A religious belief that the universe and its life forms came into existence as a divine act, creation by God.

Dictyostelium discoideum A soil-dwelling organism commonly known as slime mold that spends part of its life as a single-celled amoeba, another part as a mobile slug, and a third part as a fruiting body that produces spores.

dissipative structure Regions of organized behavior that occur in systems far from thermodynamic equilibrium and arise from the dissipation of energy.

DNA Deoxyribonucleic acid, a molecule composed of two polynucleotide chains that form a double helix and carry fundamental information in the form of genes.

emergence Properties and behaviors that arise only at the level of the full system, not for the parts of which it is made, and that exist because of interactions between those parts.

entropy In thermodynamics, entropy is the degree of disorder and appears in formulas relating free energy, enthalpy and temperature. The entropy concept is also used in other fields, such as information theory.

enzyme Proteins that catalyze biochemical reactions such as those that occur in metabolism.

equilibrium A state that does not change over time.

feedback A cause-and-effect loop in which the output is fed back in as input to the original unit. Feedback can occur in electronics, biochemical networks, or other settings and can be both positive (amplifying) or negative (damping).

flocking Collective, somewhat organized motion of a group of individuals such as birds, insects, or even inanimate drones.

fractal A mathematical object that is self-similar, appearing the same at different levels of resolution.

fractal dimension A fractional dimension used to define the measure, or length, of a fuzzy object such as a coastline or to characterize a self-similar object such as a strange attractor.

Golden Fleece Award Press releases by the late US Senator William Proxmire mocking what he saw as wasteful government spending, usually on scientific projects.

Hajj The annual Islamic ritual pilgrimage to Mecca in Saudi Arabia.

Heisenberg Uncertainty Principle A fundamental property of quantum systems that limits the accuracy with which the position and momentum of an object can be simultaneously known or specified. The principle exists at the quantum level, since matter at this level is both particle and wave.

horseradish peroxidase (HRP) An enzyme found in the root of the horseradish plant that, in its native form, catalyzes the production of lignin in the plant. The enzyme can be extracted and is used commercially to catalyze many other biochemical processes.

intelligent design A form of creationism in which its proponents argue the existence of complexity in the universe is proof of a divine creator.

labyrinth This term is sometimes used as a synonym for maze, but in this book, a labyrinth is an elaborate design usually carved into or painted on a floor that involves a single pathway into the center and no decision points along the path.

limit cycle A type of attractor associated with periodic or oscillatory behavior. The term "limit" refers to the fact that trajectories spiral toward it, reaching this cyclic limit after an infinite time has elapsed.

linga Also referred to as "lingam", this is a short, cylindrical, pillar-like object treated as a devotional image of the Hindu deity, Shiva.

lipid A type of molecule found in living systems that has fat-like qualities and is repelled by water. Lipids form structures known as bilayers, which form, in turn, cell membranes.

Lorenz model A system of three simple ordinary differential equations originally proposed to describe convection in the atmosphere. The three variables (x, y, and z) are the rate of convection, the horizontal temperature variation, and the vertical temperature variation. For some parameter values, this model exhibits chaotic behavior.

Mandelbrot set The set of complex numbers c for which the quadratic map, $z_{n+1} = z_n^2 + c$, does not diverge for all positive values of n. Images of the Mandelbrot set are usually color-coded visualizations of the boundary of the set. It exhibits self-similarity and is a fractal curve.

metaphor A figure of speech in which one thing is described as being another. A metaphor has two parts: tenor and vehicle. The vehicle is the object whose attributes are borrowed, while the tenor is the subject to which these attributes are ascribed. In "Juliet is the sun", Juliet is the tenor and the sun is the vehicle.

monad A term whose literal meaning is "unit". Although the word has found its way into philosophy, mathematics, and many other fields, Leibniz considered it to mean an indivisible unit, such as an atom.

multistability A property of a complex system in which more than one stable state exists for a given set of parameter values. A special case is bistability, in which two stable states exist.

murmuration Another term for flocking or swarming when observed in large flocks of birds, especially starlings.

mystical experience A realization with one's whole being that all things are one, interconnected, and that everything that is cannot be divided into self and non-self.

mysticism Can refer either to the study of mystical experiences or to religious and spiritual practices that emphasize and cultivate these experiences.

neuron A type of cell in nervous tissues such as the brain and spinal cord. Neurons are electrically excitable and produce spikes known as action potentials that can be detected by other nearby cells through connections known as synapses.

Nicholas of Cusa A German theologian, mathematician and astronomer who wrote about mystical experiences. He was appointed cardinal in 1448 but his work was controversial and opposed by those in authority. He was eventually excommunicated by Pope Pius II in 1460.

nonlinear dynamics A set of mathematical procedures used to study how variables change over time in systems described by nonlinear equations of motion.

oscillating chemical reaction Also sometimes referred to as chemical oscillators, these are mixtures of chemically reacting substances that exhibit periodic or oscillatory behavior. Examples discussed in this book include the Belousov-Zhabotinsky (BZ) reaction and the peroxidase-oxidase (PO) reaction.

pandemic An epidemic that has spread to engulf multiple continents or even the whole world.

pantheism A belief that all of reality and everything that exists is divine.

panentheism A belief that the divine pervades all of reality, but also goes beyond it, even beyond space and time.

period-doubling cascade A sequence of bifurcations, each of which doubles the period of oscillation. The logistic map, given by the iterative formula $x_{n+1} = r x_n(1-x_n)$, behaves this way for values of r between 3 and 4.

periodic A system that varies in time but returns to its original state after a certain period of time, then repeats the process, again and again, always returning to the same state after the same period. An example is a clock dial, where the big hand returns to its starting location after a period of one hour.

peroxidase-oxidase (PO) reaction The oxidation of organic electron donors by molecular oxygen, O_2, catalyzed by the horseradish peroxidase (HRP) enzyme. When this reaction takes place in a flow system with reduced nicotinamide adenine dinucleotide (NADH) as the electron donor, the concentrations of reactants (O_2 and NADH), as well as some enzyme intermediates, are observed to oscillate.

phase space A space, spanned by axes, in which all possible states of a system can be represented, with each point in the space corresponding to a unique state. In mechanical systems, the axes are generally taken to be positions and momentum variables.

primordial soup Life is thought to have arisen on Earth about 4 billion years ago from this mixture of inanimate substances.

protein Large polymer molecules consisting of chains of amino acids that when folded into a variety of shapes and configurations carry out many vital molecular processes in living systems. Enzymes are a type of protein, but there are many others.

predator-prey system A group of equations that describe the interactions between two competing species, a predator and its prey. The Lotka-Volterra model is the most widely studied example of a predator-prey system.

puja A ritual of worship in Hinduism focused on reverence and paying homage to the divine, usually a specific deity. It can be performed in a temple, the home, on special occasions, or even daily. One step in a puja ceremony can involve circumambulation around a shrine to the deity.

quantum mechanics A system of equations and theories that describe the motions and energies of microscopic systems at the atomic or subatomic level. One of the key features of the quantum mechanical formalism is the duality of matter as both wave and particle.

Rössler model A system of three nonlinear ordinary differential equations proposed by Otto Rössler as a model for a chemical oscillator. For certain parameter values, the model exhibits chaotic behavior.

second law The second law of thermodynamics defines the direction in which processes occur spontaneously, by specifying that they always occur in such a way that entropy increases. Note that other second laws exist, such as the second law of motion attributed to Newton, namely $F = ma$.

self-organization The spontaneous appearance of order in a system arising from nonlinear interactions in a system that is far from thermodynamic equilibrium. Examples include limit cycle oscillations, pattern formation, spontaneous wave behavior, and chaos.

self-organized criticality (SOC) A critical point that emerges as an attractor in certain nonlinear systems, such as a sand pile.

self-similarity A property of a system, such as a fractal, in which the parts are exactly or approximately the same as the whole.

shared space A design method for traffic in urban settings that eliminates most signage, lane markings and traffic control lights, relying instead on the tendency of crowds to self-organize their flow.

slime mold A eukaryotic organism, usually found in the soil or on forest floors, that lives part of its life cycle as a single-celled organism and other parts aggregating to form a mobile slug and, eventually, a fruiting body. One extensively studied form is *Dictyostelium discoideum*.

spruce budworm The larval, caterpillar form of the moth *Choristoneura*. The caterpillars are known to kill large numbers of coniferous trees, particularly in the western United States.

state space A type of Euclidean space in which the axes are variables that define the state.

strange attractor An attractor that has a fractal structure. These are usually associated with chaotic dynamics.

stupa A mound-shaped structure containing relics, usually some sort of remains of a revered figure, that is used as a place of meditation or ritual. Circumambulation of the stupa is a common ritual.

sutra A Sanskrit word for scripture; a teaching or set of rules.

synapse The connection between neurons where electrical connections and chemical species known as neurotransmitters carry signals from one cell to another.

systems theory A form of study that emphasizes the connections and interactions between the parts that make up a system.

Tathagata In the Buddhist tradition, this refers to one who has transcended the human condition and come to know the essential truth: the Buddha.

thermodynamics A system of equations and theories that describe the behavior and interrelationships of heat, energy and work.

trajectory The path followed by an object or system moving under the influence of some force. The path can be in physical space (Cartesian space), phase space or state space.

Twitter Revolution Originally used to describe the protests and demonstrations that erupted in 2009 following the Iranian national elections, so-called because protestors kept in contact with each other using Twitter.

universality Generally refers to the similarity of phenomena or mathematical descriptions applicable to a wide array of physical systems or examples.

Further Reading

Ralph Abraham and Christopher Shaw, *Dynamics: The Geometry of Behavior (Parts 1–4)*, Aerial Press, Santa Cruz, California (1985)

Tsultrim Allione, *Women of Wisdom*, Snow Lion, Revised & Enlarged Edition, New York (2000)

Karen Armstrong, *A History of God*, Knopf, New York (1993)

Karen Armstrong, *Visions of God: Four Medieval Mystics and their Writings*, Bantam Books, New York (1994)

Lauren Artress, *Walking a Sacred Path: Rediscovering the Labyrinth as a Spiritual Tool*, Riverhead Books, New York (1995)

Anne Bancroft, *Weavers of Wisdom: Women Mystics of the Twentieth Century*, Penguin Books, London (1989)

Ian G. Barbour, *Religion and Science*, HarperOne Revised, San Francisco (2013)

Michael Barnsley, *Fractals Everywhere*, Academic Press, New York (1988)

Gregory Bateson and Mary Catherine Bateson, *Angels Fear*, MacMillan, New York (1987)

David Bohm and F. David Peat, *Science, Order and Creativity*, Bantam Books, New York (1987)

John Briggs and F. David Peat, *Seven Life Lessons of Chaos: Spiritual Wisdom from the Science of Change*, HarperCollins, San Francisco (2000)

Bonaventure, *Bonaventure: The Soul's Journey into Life; The Tree of Life; The Life of St. Francis*, Paulist Press, New York (1982)

Bruno Borchert, *Mysticism: Its History and Challenge*, Samuel Weiser, Inc., York Beach, Maine (1994)

Roland Bouffanais, *Design and Control of Swarm Dynamics*, Springer Briefs in Complexity, Springer, Singapore (2016)

Joseph Campbell, *Myths to Live By*, Penguin Books, New York (1972)

Pierre Teilhard de Chardin, *The Phenomenon of Man*, Harper and Row, New York (1959)

Francis S. Collins, *The Language of God: A Scientist Presents Evidence for Belief*, Free Press, New York (2006)

Kenneth Davis, "The Quest for the Negentropic Grail," *Earthrise Newsletter*, Vol. 2, No. 4 (1974)

Richard Dawkins, *The God Delusion*, Houghton Mifflin Harcourt, Boston (2006)

R.L. Devaney, *A First Course in Chaotic Dynamical Systems: Theory and Experiment*, Addison-Wesley, Massachusetts (1992)

Diana L. Eck, *Encountering God: A Spiritual Journey from Bozeman to Banaras*, Beacon Press, Boston (1993)

Elaine Howard Ecklund, *Science vs. Religion: What Scientists Really Think*, Oxford University Press, New York (2010)

Arthur S. Eddington, *Science and the Unseen World*, MacMillan, New York (1929)

Taner Edis, *The Ghost in the Universe: God in Light of Modern Science*, Prometheus Books, New York (2002)

Irving R. Epstein and John A. Pojman, *An Introduction to Nonlinear Chemical Dynamics: Oscillations, Waves, Pattern and Chaos*, Oxford University Press, Oxford (1998)

Richard J. Field and Maria Burger, *Oscillations and Traveling Waves in Chemical Systems*, John Wiley & Sons, New York (1985)

Matthew Fox, *The Coming of the Cosmic Christ*, Harper & Row, New York (1988)

Richard Gallagher and Tim Appenzeller, "Complex Systems — Special Issue," *Science*, Vol. 284, Issue 5411, pp. 79–109 (1999)

Langdon B. Gilkey, *Nature, Reality and the Sacred: The Nexus of Science and Religion*, Fortress Press, Minnesota (2000)

Malcolm Gladwell, *The Tipping Point: How Little Things Can Make a Big Difference*, Little Brown, New York (2000)

Leon Glass and Michael Mackey, *From Clocks to Chaos: The Rhythms of Life*, Princeton University Press, New Jersey (1988)

James Gleick, *Chaos: Making a New Science*, Viking, New York (1987)

Albert Goldbeter, *Biochemical Oscillations and Cellular Rhythms: The Molecular Bases of Periodic and Chaotic Behavior*, Cambridge University Press, Cambridge (1996)

L.O. Gomez, "The Whole Universe as a Sutra," *Buddhism in Practice*, D. S. Lopez, ed., pp. 107–112, Princeton University Press, New Jersey (1995)

Al Gore, *Earth in the Balance: Ecology and the Human Spirit*, Plume, New York (1993)

Stephen Jay Gould, *Eight Little Piggies*, W.W. Norton & Co., New York (1993)

Barbara Bradley Haggerty, *Fingerprints of God: The Search for the Science of Spirituality*, Riverhead Books, New York (2009)

H. Haken, *Synergetics, An Introduction: Nonequilibrium Phase Transitions and Self-Organization in Physics, Chemistry and Biology*, Springer-Verlag, Berlin (1983)

John Haught, *A John Haught Reader: Essential Writings on Science and Faith*, Wipf and Stock, New York (2018)

Dirk Helbing et al., "Improving Pilgrim Safety During the Hajj: An Analytical and Operational Research Approach," *Interfaces*, Vol. 46, No. 1, pp. 74–90 (2016)

John H. Holland, *Emergence — From Order to Chaos*, Addison-Wesley, New York (1998)

Richard H. Jones, *Science and Mysticism: A Comparative Study of Western Natural Science, Theravada Buddhism and Advaita Vedanta*, Bucknell University Press, Lewisburg (1986)

Carl G. Jung and Wolfgang Pauli, *The Interpretation of Nature and the Psyche*, Pantheon, New York (1955)

Stuart Kauffmann, *The Origins of Order: Self-Organization and Selection in Evolution*, Oxford University Press, Oxford (1993)

Stuart Kauffmann, *Reinventing the Sacred: A New View of Science, Reason and Religion*, Basic Books, New York (2008)

Theresa King, ed., *The Spiral Path: Exploration's in Women's Spirituality*, Y.E.S. International Publishers, Minnesota (1992)

George Lakoff and Mark Johnson, *Metaphors We Live By*, University of Chicago Press, Chicago (1980)

Raima Larter, "Understanding Complexity in Biophysical Chemistry," *Journal of Physical Chemistry*, Vol. 107, pp. 415–429 (2003)

Raima Larter, Lars Folke Olsen, Curtis G. Steinmetz and Torben Geest, "Chaos in Biochemical Systems: The Peroxidase Reaction as a Case Study," in *Chaos in Chemistry and Biochemistry*, Richard J. Field and Laszlo Gyorgyi, eds., World Scientific, Singapore (1993), pp. 175–224.

Raima Larter, "Life Lessons from the Newest Science: Bifurcation," *Noetic Sciences Review*, No. 59, pp. 22–27 (2002)

"Better Living through Chaos," *The Economist*, Vol. 352, No. 8137, pp. 89–90 (1999)

Bentley Layton, *The Gnostic Scriptures*, Doubleday & Co., New York (1987)

George Leonard, *"The Silent Pulse: A Search for the Perfect Rhythm that Exists in Each of Us,"* E.P. Dutton, New York (1978)

Stephen Levine, *Guided Meditations, Explorations, and Healings*, Doubleday, New York (1991)

Edward N. Lorenz, "Deterministic Nonperiodic Flow," *Journal of Atmospheric Science*, Vol. 20, p. 130 (1963)

Edward N. Lorenz, *The Essence of Chaos*, University of Washington Press, Washington (1993)

Alfred J. Lotka, *Elements of Mathematical Biology*, Dover Publications, New York (1956)

Joanna Macy, *Despair and Power in the Nuclear Age*, New Society Publishers, Philadelphia (1983)

Benoit B. Mandelbrot, *The Fractal Geometry of Nature*, W. H. Freeman and Co., San Francisco (1983)

Alistair E. McGrath, *Science and Religion: An Introduction*, 3rd Edition, Wiley-Blackwell Publishing, New Jersey (2020)

Stanley L. Miller, "A Production of Amino Acids under Possible Primitive Earth Conditions", *Science*, Vol. 117, Issue 3046, pp. 528–529 (1953)

Walter Moore, *A Life of Erwin Schrödinger*, Cambridge University Press, Cambridge (1994)

Harold Morowitz, *The Emergence of Everything: How the World Became Complex*, Oxford University Press, New York (2002)

Erik Mosekilde and Ole Mouritsen, eds., *Modeling the Dynamics of Biological Systems: Nonlinear Phenomena and Pattern Formation*, Synergetics, Springer-Verlag, Berlin (1995)

Tom Mullin, ed., *The Nature of Chaos*, Clarendon Press, Oxford (1993)

James D. Murray, *Mathematical Biology*, Springer-Verlag, New York (1990)

Gregoire Nicolis, *Introduction to Nonlinear Science*, Cambridge University Press, Cambridge (1995)

Gregoire Nicolis and Ilya Prigogine, *Self-Organization In Nonequilibrium Systems: From Dissipative Structures to Order through Fluctuations*, John Wiley & Sons, New York (1977)

Gregoire Nicolis and Ilya Prigogine, *Exploring Complexity: An Introduction*, W. H. Freeman & Co., New York (1989)

Kees van Kooten Niekerk and Hans Buhl, eds., *The Significance of Complexity: Approaching a Complex World through Science, Theology and the Humanities*, 1st Edition, Routledge, London (2020)

E. Ott, *Chaos in Dynamical Systems*, Cambridge University Press, Cambridge (1993)

Willard G. Oxtoby and Roy C. Amore, *World Religions: Eastern Traditions*, 3rd Edition, Oxford University Press, New York (2010)

Parker Palmer, *To Know as We are Known: Education as a Spiritual Journey*, Harper San Francisco (1993)

G. Parrinder, *Mysticism in the World's Religions*, Oxford University Press, New York (1976)

M. Scott Peck, *A World Waiting to be Born*, Bantam Books, New York (1993)

John Polkinghorne, *Science and Religion in Quest of Truth*, Yale University Press, Connecticut (2011)

S. Prabhavananda and F. Manchester, *The Upanishads*, Penguin Putnam, New York (1948)

Ilya Prigogine and Isabelle Stengers, *Order out of Chaos: Man's New Dialogue with Nature*, Bantam Books, New York (1984)

Ilya Prigogine, *The End of Certainty*, Simon & Schuster, New York (1997)

Kathleen Raine, *Farewell Happy Fields*, Hamish Hamilton, London (1974)

Rainer Maria Rilke, *Rilke's Book of Hours: Love Poems to God,* translated by Anita Barrows and Joanna Macy, 100th Anniversary Edition, Riverhead Books, New York (2005)

D. Rosen, *The Tao of Jung: The Way of Integrity*, Viking Penguin, New York (1996)

Otto E. Rössler, "An Equation for Continuous Chaos," *Physics Letters*, Vol. 57A, No. 5, pp. 397–398 (1976)

Robert John Russell and Joshua M. Moritz, eds., *God's Providence and Randomness in Nature: Scientific and Theological Perspectives*, Templeton Press, New York (2019)

Erwin Schrödinger, *What is Life?*, Cambridge University Press, Cambridge (1967)

S.S. Schweber, "Physics, Community and the Crisis in Physical Theory," *Physics Today*, pp. 34–40 (1983)

B. Speelman, R. Larter and R.M. Worth, "A Coupled ODE Lattice Model for the Simulation of Epileptic Seizures," *Chaos*, Vol. 9, pp. 795–804 (1999)

Jonathan Star and Shahram Shiva, translators, *A Garden Beyond Paradise: The Mystical Poetry of Rumi*, Bantam Books, New York (1992)

Victor Stenger, *Has Science Found God? The Latest Results in the Search for Purpose in the Universe*, Prometheus Books, New York (2003)

Mikael Stenmark, *Rationality in Science, Religion and Everyday Life*, University of Notre Dame Press, Indiana (2016)

Ian Stewart, *Does God Play Dice? The Mathematics of Chaos*, Blackwell, Massachusetts (1989)

Steven Strogatz, *Nonlinear Dynamics and Chaos*, Addison-Wesley, Massachusetts (1994)

Steven Strogatz, *Sync: The Emerging Science of Spontaneous Order*, Hachette Books, London (2003)

Mariusz Tabaczek, *Emergence: Toward a New Metaphysics and Philosophy of Science*, University of Notre Dame Press, Indiana (2019)

Phyllis Tickle, *The Great Emergence: How Christianity is Changing and Why*, Baker Books, Michigan (2012)

Anne B. Ulanov, *Primary Speech: A Psychology of Prayer*, John Knox Press, Atlanta (1982)

Evelyn Underhill, *Mysticism: The Nature and Development of Spiritual Consciousness*, 12th Edition, Oneworld Publications, London (1993)

Vito Volterra, "Variations and Fluctuations of a Number of Individuals of Animal Species Living Together (Translated from the Original Italian, 1926)," in *Animal Ecology*, R.N. Chapman, ed., McGraw-Hill, New York (1931), pp. 409–448.

Mitch Waldrop, *Complexity: The Emerging Science at the Edge of Order and Chaos*, Simon & Schuster, New York (1992)

Alan Watts, *Behold the Spirit: A Study in the Necessity of Mystical Religion*, Random House, New York (1971, reprint of 1947 book)

Simone Weil, *Waiting on God*, Routledge and Kegan Paul, London (1951)

Simone Weil, *Gravity and Grace*, Routledge and Kegan Paul, London (1952)

Simone Weil, *Love in the Void: Where God Finds Us*, Plough Publishing House, London (2018)

Steven Weinberg, *The First Three Minutes*, Basic Books, New York (2020)

Margaret J. Wheatley, *Leadership and the New Science: Learning about Organization from an Orderly Universe*, Bernett-Koehler, San Francisco (1992)

Arthur Winfree, *The Geometry of Biological Time*, Springer, New York (1980)

F.E. Yates, ed., *Self-Organizing Systems: The Emergence of Order*, Plenum Press, New York (1987)

James Francis Yockey, *Meditations with Nicholas of Cusa*, Bear and Company, New Mexico (1987)

Louise B. Young, *The Unfinished Universe*, Oxford University Press, Oxford (1993)

Photo & Illustration Credits

Use of the following photos and illustrations is gratefully acknowledged. All photos and illustrations not listed below are the author's own work.

Cover Image The cover image is a composite including one photo of the Horsehead Nebula taken by the Hubble telescope. This photo is in the public domain. Credit: NASA, ESA, and the Hubble Heritage Team (STSci/AURA) https://www.nasa.gov/mission_pages/hubble/science/horsehead-different.html

Figure 2 Credit: Adi Wahid; Creative Commons Attribution-Share Alike License, Wikimedia: https://commons.wikimedia.org/wiki/File:The_Kaaba_during_Hajj.jpg

Figure 4 Credit: Mark R. Leach, *Chemogenesis: Chemical Systems Prone to Complexity*, a free, open-source webbook available at: https://www.meta-synthesis.com/webbook/24_complexity/complexity3.php

Figure 5 Credit: Cornelis Weijer, University of Dundee.

Figure 8 Credit: Sheryl Hemkin, PhD thesis, Purdue University.

Figure 9 Credit: James D. Murray, *Mathematical Biology*, Springer-Verlag, New York (1990), p. 84.

Figure 11 Credit: Lars Folke Olsen.

Figure 12 Credit: Dschwen; Creative Commons Attribution-Share Alike License, Wikimedia: https://en.wikipedia.org/wiki/File:Lorenz_attractor_yb.svg

Figure 13 Credit: Lauren Artress, *Walking a Sacred Path: Rediscovering the Labyrinth as a Sacred Tool*, Riverhead Books, New York (1995)

Figure 14 Credit: Wolfgang Beyer; Creative Commons Attribution-Share Alike License, Wikimedia: https://commons.wikimedia.org/wiki/File:Mandel_zoom_00_mandelbrot_set.jpg

Figure 19 Credit: User Wxs; Creative Commons Attribution-Share Alike License, Wikimedia: https://commons.wikimedia.org/wiki/File:KochFlake.svg

Figure 20 Credit: Benjamin Wargo and Norman Garrick, University of Connecticut.   https://www.cnu.org/publicsquare/shared-space-intersections-mean-less-delay

Figure 21 Credit: Brent Speelman, PhD thesis, Purdue University.

Figure 22 Credit: Sheryl Hemkin, PhD thesis, Purdue University.

Figure 23 Credit: Brent Speelman, PhD thesis, Purdue University.

Index

adaptive behavior, 85
adaptive systems, 115, 116, 119
Anderson, Philip, 116
Apollo astronauts, 88
Arab Spring, 128, 129
Artress, Lauren, 89
attractor, 1, 5–16, 18, 19, 21–31, 54,
  58, 60, 64, 65, 73, 74, 78, 79, 86–88,
  92–94, 96, 97, 105, 108, 110, 111, 125
Augustine, 105
autocatalysis, 65
autocatalytic reaction, 65
Avatamsaka, 106

Belousov-Zhabotinsky (BZ) reaction, 34,
  35, 42–44, 46, 56–61
bifurcate, 4, 7, 9, 10, 16,18–25, 27, 30, 54,
  60, 67, 70, 73, 77–79, 86, 87, 92, 110
bistability, 64
Bohm, David, 103
Bohr, Niels, 115
Boltzmann constant, 116
Borchert, 106, 107
breakthrough, 110
breath, 70, 71

Bruno, Giordano, 105, 106
budworm, 23, 24, 64
butterfly effect, 83, 86

calcium, 63, 64, 67, 72
cardiac pacemaker cells, 61, 63, 67
carrying capacity, 23, 24, 64
cell division cycle, 61
center, 11, 13–15, 28–30, 88–92, 97, 110,
  111
chaos, 3, 4, 8, 15, 16, 28, 33, 74, 75,
  77–87, 89, 92–94, 96, 98, 113, 116, 122,
  123, 125, 126
chaotic, 4, 8, 16, 28, 77, 79, 80, 82–88,
  92–94, 96–98
Chartres, 88–90
chemical oscillation, 58, 63
chemical reaction, 4, 5, 34, 36, 42,
  58–60, 65, 73, 79, 84
Cicada, 62, 63, 72
circadian rhythms, 61
circumambulation, 11–13, 88, 92
classical mechanics, 7
climate change, 53, 132
colonies, 48, 49

Acknowledgements

This book has been in the works for almost 30 years, so it is especially difficult to acknowledge all those who have helped and encouraged me along the way. But I will give it a try and hope I don't leave anyone out.

Early drafts of this manuscript were read and commented on by numerous former colleagues at IUPUI including Ken Davis, Bill Jackson, Rick Ward, Barbara Cambridge, Sharon Hamilton, Carol Parish and Stan Sunderwirth. Wise and generous friends also read and provided feedback, including Melba Hopper, Jean Denton, Annie Carpenter, Denise Deig, Kim Watanabe, Carol Montgomery, Cathy Cobb and George Leonard.

My writing group deserves special mention, since some of these folks read more than one version, and provided valuable insights. Thanks to Hildie Block, Pragna Soni, Margaret Rodenberg, Lorie Brush, Stephanie Joyce, Susan Lynch, the late Melanie Otto and Christine Jackson for reading and providing valuable feedback on early versions of the manuscript. My former literary agent, the late Loretta Barrett, also read the early manuscript version with a keen eye and provided lots of helpful insights and advice. My current writing group provides a seemingly endless source of inspiration and moral support. Thanks to Tara Campbell, Marcy Dilworth, Myna Chang, Mary Sophie Filicetti, Jessie Seigel, Beth Wenger and Raegan O'Lone.

I also need to thank those colleagues, some of whom are mentioned in this book, for their contributions to our joint work, their friendly collaboration on numerous projects, and their good will. Special thanks to Alex Scheeline, Ken Showalter, Lars Folke Olsen, the late Baltz Aguda, Curt Steinmetz, Sheryl Hemkin, Bob Worth, Brent Speelman, and Torben Geest.

Big thanks to those people who have helped me learn how to write about science for the public, including my instructors and fellow students at Johns Hopkins University and the Writers Center in Bethesda, especially David Taylor, Mary Knudson and Nancy Shute. Thank you also to my colleagues and editors at the American Institute of Physics, including Larry Frum, Wendy Beatty, Emilie Lorditch, Julia Feynman and numerous others who have been a joy to work with in recent years.

Thanks also to my editor, Shaun Tan Yi Jie at World Scientific, for inviting me to finish and publish this book and providing invaluable help along the way. This book would truly not exist without your help and support.

Finally, thank you to my family, who were always there as I struggled to bring this book to life. To my sons, Nathan and Ben, and my husband, Ken: this one's for you!

About the Author

Dr. Raima Larter retired as Professor of Chemistry at Indiana University-Purdue University at Indianapolis (IUPUI), USA, in 2003 and moved to Washington to direct the country's research program in theoretical and computational chemistry at the National Science Foundation (NSF) for over ten years. In addition to three decades of research and teaching at IUPUI, she also served in a variety of administrative positions, including Associate Dean, Department Chair and President of the Faculty. She taught courses on topics from freshman chemistry to advanced undergraduate and graduate courses in thermodynamics, kinetics and statistical mechanics. Her research interests were and continue to be in complex systems, especially in biophysical settings. She holds a PhD in Chemistry from Indiana University, USA.

While at NSF, Dr. Larter also edited the agency's newsletter, *The Current*, for a few years which kickstarted her foray into writing about science for the public. At the same time, she continued to pursue her work in complex systems science through agency initiatives and became especially interested in the application of complexity and emergence ideas to social systems. She obtained an MA in Writing from Johns Hopkins University in 2016 and is now a full-time writer.

Dr. Larter currently does freelance science writing for a number of organizations, including the American Institute for Cancer Research, National Science

Foundation, American Chemical Society, and American Institute of Physics. She has published numerous short stories and essays, some of which have won awards. She is a Pushcart Prize nominee and the author of two fiction novels, *Fearless* and *Belle o' the Waters*, of which the former was a finalist in the 2017 Faulkner-Wisdom Fiction Contest. She hosts the blog *Complexity Simplified* (http://raimalarter.com), where she discusses topics ranging from science to spirituality and writing.

Printed in the United States
by Baker & Taylor Publisher Services